Praise for *All Through t*

A *Forbes* Best New Books on Space 2024

'Rarely is a non-fiction book about science this engaging.'

—*Forbes*

'Passion and urgency lie beneath poetic and whimsically written passages.'

—*Sky at Night Magazine*

'Roberston is fantastic when she's angry; when she gets earnest ... This is powerful stuff.'

—Niall Griffiths, *Nation Cyrmu*

'A heartfelt, necessary and very enjoyable book.'

—Tristan Gooley

'A hymn of praise to darkness and the unfathomable wonder of a true night sky, this book is also an urgent call to arms. As Dani Robertson shows, our health, and that of the planet around us, is inextricably linked with the power of the dark. We are losing it at great speed, and to our great peril. Read the book, look up in awe, and act.'

—Mike Parker

'An utterly illuminating book that will open your eyes to an overlooked world in deep peril. Dani makes a compelling case for just how urgently we need to reform our relationship with darkness. Everyone interested in nature will find surprise, intrigue and awe on every page.'

—Nicholas Gates

'To read *All Through the Night* is to experience stars appearing one by one in the night sky; discreet, glowing insights throw gentle but piercing light onto what we are doing to what Dani Roberston calls one of the most endangered landscapes on Earth – the night sky. The civilised world has flooded the velvety blackness with strange glowings-on that disorientate insects, birds and sealife. We flood our own psyches with permanent light that shields us from what we have evolved to do in the dark hours – rest and reflect. I loved being lost in the real stuff of darkness, hearing the Shearwaters and owls claim their territory, and was disturbed by how we have, often unintentionally, disrupted their worlds. Even the trees suffer from our outpouring of artificial light, as does the vastness of space itself as satellites march across the heavens. The book ends on what we can do to embrace darkness again, to welcome it back and reduce our light pollution. It is a thought-provoking book that sheds light onto our need to love darkness.'

—Mary Colwell

'It's rare to read something so important but written in a way that it actually inspires you to shout about it from the rooftops. Important, informative, inspiring, passionate, and written, extremely beautifully, by someone who really knows her stuff. One of my favourite books so far this decade.'

—Brigit Strawbridge

'Dani Robertson is a fine writer – lyrical and eloquent in extolling the beauty of our world – especially the dark starry sky. The vault of heaven' has inspired mystery and wonder since the dawn of humanity but is now threatened and despoiled – like so much of nature – by ill-applied modern technologies. She recounts reminiscences of a dedicated life campaigning for conservation of our natural world. But her book is more than a memoir; it's filled with episodes enlivening her theme with history, science and topography, and with individual biographies. *All Through the Night* is fascinating and inspiring – it deserves wide readership.'

—Lord Martin Rees, Astronomer Royal

'Woven through *All Through the Night* is a wonderful story of what natural darkness means to Dani. Her story is more than an anecdote, it's a powerful recounting of what is lost when the night burns like day, and the simple steps we can all take to reclaim natural darkness to benefit us all.'

—Ruskin Hartley, Executive Director at
International Dark-Sky Association

Dani Robertson, a Dark Sky Officer for Eryri National Park and the Areas of Outstanding Natural Beauty, is originally from Greater Manchester but moved to the Welsh countryside at an early age. She is prolific in conservation work, championing the darkness for all, and is a regular speaker at public outreach events. Her advocacy for night skies was recognised by the International Dark Sky Association in 2022, when she received the Dark Sky Defender Award. *All Through the Night* is her first book and will equip readers with the tools for defending our skies. She lives in Cymru with her husband and their two dogs and two cats.

ALL THROUGH THE NIGHT

ALL THROUGH THE NIGHT

HOW DARK SKIES CAN SAVE OUR WORLD

DANI ROBERTSON

HarperNorth
Windmill Green,
Mount Street,
Manchester, M2 3NX

A division of
HarperCollins*Publishers*
1 London Bridge Street
London SE1 9GF

www.harpercollins.co.uk

HarperCollins*Publishers*
Macken House, 39/40 Mayor Street Upper
Dublin 1, D01 C9W8, Ireland

First published by HarperNorth in 2023

1 3 5 7 9 10 8 6 4 2

A catalogue record for this book
is available from the British Library

PB ISBN: 978-0-00-858675-1

Printed and bound in the UK using 100% renewable electricity at
CPI Group (UK) Ltd

This book is produced from independently certified FSC™ paper
to ensure responsible forest management.

For more information visit: www.harpercollins.co.uk/green

To all the creatures of the night and
those who fight against the light.

Contents

Author's Note

What if I told you I knew one of the ways to save the world?

What if I told you that natural darkness is under threat? What if I told you that the night is practically endangered? Would you think twice before switching on the lights?

What would you think if I told you that it was as simple as flicking a switch?

Would you do it?

Every evening, around the world, each one of us heads home as the day fades away around us. Every evening, millions of us flick switches that illuminate our lives without giving it a second thought. Our homes burn bright all through the night, as electricity runs like rivers along our streets and roads ensuring the night has nowhere to hide and that we can walk in an artificial daylight, be it seven in the evening or one in the morning, as if the dark is powerless to reach us.

From our biggest cities to our smallest towns, light is all around us, all of the time. It is inescapable. It has barricaded us into a starless existence, with the Milky Way now completely lost to over one third of the global population. There is no other habitat we have degraded more than our darkness. Of all the ways we have robbed the natural world, nothing even holds a candle in comparison to the damage we have done to darkness.

What if I told you that in many places in the Western world darkness is already extinct? For ninety-nine per cent of us westerners, the sun set on a natural night-time decades ago. Light pollution is the issue no one has heard of, but it's the one we are all impacted by, whether we realize it or not.

We have become drunk on power, wielding lightsabers, cutting through the night sky and banishing our age-old foe: the dark. We have been too liberal, placing lights where they are not needed. Bathing our hillsides and woodlands in untamed light. Painting over rivers and coastlines with lashings of luminosity. We haven't paused to think about the darkness. What it means to us. What it means to the natural world.

We haven't even looked back, as darkness has been banished, beyond reach for everyone except those with the privilege of travelling to Dark Sky places and starry night skies, such as countries found deep in the Northern Hemisphere.

We are all born in the darkness, moulded in it, and it is into the darkness that we will all return. Our formative experience as a living thing is existing in the warm and comforting darkness of our mother's womb. We do not fear the dark then. But from the moment we are pushed into the bright light of day, we learn to fear the dark, to avoid it, to destroy it.

But we have lost too much, too suddenly. We are quickly running out of places where darkness still exists. We are entering a new realm of the Anthropocene, one where a natural night no longer survives. This is our last chance to see the night sky as it has existed for all of human history, before it is changed forever by man-made satellites, or hidden from human eyes by an impenetrable fortress of artificial light that imprisons us all in a perpetual, artificial existence.

It has come at a cost, greater than we realize. It is making us sick. Our health is on the line, but can we act in time?

Our wildlife is suffering, since, just like us it requires natural darkness. The up-lit ancient oak, its boughs now purple, pink and blue, was once a home to nesting birds, the insects that once nested there are no longer able to find themselves comfort in their own habitats. No squirrels hunt for acorns. Even the tree itself is suffocating from the light that clings to its canopy.

As you lie in bed you hear the delicate and desperate fluttering of a moth's wings against the landing light you left on, its fragile wings burning, turning to dust. Another victim dead before the dawn.

Like a moth to the flame, I have always been drawn to darkness.* I have peered down into the oceanic depths and wondered what world exists beyond the reaches of the sun's light. I have gazed upward to the countless filaments that stud the velveteen heavens and contemplated our very existence. I have walked through the heart of a wildwood, feet grasped by roots and brambles as they extend from the dense and dark thickets. I have sat in silence beside the sea and watched the moon rise, its light gracefully reaching the sands, awakening the life among the tideline, life that is alien to us day dwellers.

But those experiences are slipping away, fading quicker than the sun at the days end. On every horizon I see moels arising (to those who don't speak Welsh, moel describes hills being 'bald'– having an upper, rounded outline), like ghostly landforms, translucent but definitively present. Some yellow, some white, some blue. Light-pollution balloons from the city below it, a semi-permanent marker, highlighting the wasteful ways of the human life caught under it.

So that's why I work to protect the largest Dark Sky Reserve in Wales.† To go from living in one of the most

* I'm not just talking about my 'emo' phase.

† I'm like Batman but with less money and not even close to having the same amount of uniform.

heavily polluted areas of the UK to protecting some of our best Dark Skies, means that I have never taken the dark for granted. I remember what it was like to look up and see nothing but an orange haze above my head. The excitement of hearing about meteor showers and visiting comets, being replaced with disappointment (and an aching neck) as I stood on illuminated streets, separated from the sky by a barrier of light. I understand what it means to stand beneath your first star-filled night sky and be completely blown away by the depth and beauty of thousands of stars hanging in a crystalline sky, the tranquillity that washes over your mind when, finally, it is given a break from incessant incandescents and LEDs.

I know what so many have lost, so I know how important it is that we protect what is left, and restore the stars to our skies where the dark has been stolen, so suddenly. I am one of the lucky ones. I can step out of my front door and, within a few minutes of walking, break free of the illuminated border and be out into an oasis of darkness. My horizon is still plagued by domes of light here and there, but above me arcs the Milky Way, while Orion hangs above the hill, leading me out into the night.

Darkness must not become another part of our world that is sacrificed due to the human condition. It must not become another point of access for the elite, for those who can jet off to find darkness in far-flung places. Darkness should not be a privilege afforded to the few. The night sky is for everyone, all beings, all things. Solving light pollution will not stop the climate and biodiversity crises on its own, but as you will read, it's a piece of the puzzle that we have the answers to. Right now. A jigsaw piece that is easy, simple and cheap to slot into place. It's an edge piece, with two flat edges, guiding us to easily do the right thing, buying us much needed time while we solve more perplexing pieces that we are not yet sure how to piece together. So, why aren't we acting?

Where is the rage against the dying of the night?

CHAPTER ONE

Who's Afraid of the Dark?

Imagine your favourite beautiful place.

Sunlight filters through, warming your skin.

Perhaps the coconut-esque scent of gorse fills your nose, and birdsong fills your ears.

Now, imagine that place being obscured by a haze, a vapour, a mist. Your eyes obscured by a gloomy fog.

You can no longer see the flowers, nor the river's winding bank and the gently swaying branches of the willowy trees. The sunlight that once invigorated you has been blocked off.

Its rays cut away.

You feel disorientated.

Which way did you walk here?

How will you get back?

The landscape and everything in it has been smothered.

Now, imagine realizing this veil is a pollutant. A pollutant that could easily be controlled, yet it has been left to meander its way over everything; blanketing everyone, removing senses, and choking the life out of the living things beneath it. We wouldn't let this happen to the daylight. So, why have we let it happen to our darkness? Some say we can't see the wood for the trees, but in this case we can't see the glow for the glare.

You might be afraid of the dark. In that case, this book is *definitely* for you.

I am not here to sneer or judge those of you who fear the dark, nor to make you feel foolish for fearing the shadows. To be scared of the dark is completely human. It's the reason your DNA has survived through to its existence in you. This fear is intrinsically human.

I am not here to demand that all lights around the globe be switched off and plunge us all into an unlit night.

For too long, darkness has been wrongly blamed for societal issues and used as a symbol for all things evil. I am here to show you the way to enlightenment in the night, to help you find your place in the darkness, and to show you how our bodies and minds need the relief of natural night; how our wildlife suffers from our human addiction to light. I will show you how to take control of the blight that illuminates our homes so brightly that we can see the glow from space.

As a young girl, I vividly remember the brightly lit streets of Greater Manchester. The grit of the streets and the dust from the roads that caught in my eyes and made them cry. The dark seemed so terrifying then; so many powerful lights blinded me further, casting deep and dangerous shadows all around. I would cling to my dad's hand and keep myself as close as possible while we walked those sodium-lit streets, every shadow exaggerated through poorly planned-for lighting. There was never any hope of seeing a shooting star. Stars were reserved for fairy-tales and nursery rhymes – and the faint, luminescent green of the plastic constellations stuck on my bedroom wall.

I will never forget my first Dark Sky. Like most people of my generation, I had never experienced anything close to true darkness before. I had never stood beneath the universe to witness a thousand stars shining back, the ancient lights of human fascination. I didn't get to hear the stories in the skies, for they were nowhere to be seen; the stage for the greatest show on Earth hidden behind a curtain of light. Too many of us in the West and beyond share the same fate. Ninety-nine per cent of people in North America and Western Europe live under heavily polluted skies. That's nearly 600 million of us, denied a right to see the wonders of the universe.

Our greatest resource, which has inspired us to venture out from our caves and travel out to the moon and beyond, lies out of reach to almost everyone. A few of us will see a handful of stars, but most will see nothing at all – just the moon, and, if they're lucky, the brightest planets. A whole third of humanity has been severed from our home galaxy – the Milky Way. It is a tragedy that so many of us have been cut off from this ancient source of inspiration for the most creative and scientific minds of our species. *Who are we to deny them their opportunity to flourish by preventing them this basic right? What bright mind and free-thinking genius have we imprisoned in the glare of lights?*

I want you to fall in love with the night. I want you to embrace the darkness, to care for it, and for all the things that exist within it. I want you to protect it and fight for it before it is gone forever – to realize how much we have lost already, and what we risk with our endless creation of light. Most of all, I want you to understand and respect darkness, and to find your place within it.

I have used research and science from some of the most fantastic people on the planet, who are trying to sound the alarm. You will find the evidence is stacking up on the side of the night. Some areas in this field (like satellites) are advancing so quickly that soon after I write this sentence, the situation will have changed. The pace of these changes is coming at the speed of light. So much is happening, that most of us won't even realize what's happened until it's too late.

But there's a glimmer of hope regarding light pollution. It doesn't have to be this way. We have the knowledge to fix this. If we can bring together the voices of communities and empower them to call for action, our politicians and governments will know we want to fight to right this wrong that has befallen the starry night.

With this book, I want you to feel empowered to make a simple, but impactful change that has the potential to ripple out from you and travel all the way to space. It's like 'darkness dominoes'. Start the chain reaction in your own home, your own community. You have got the power to make real change.

My first glimmer of starlight changed my life forever. I was never the same again. I have spent years seeking out darkness, in places old and new. Although terrifying at first, I feel safer in the shroud of darkness than I ever did in the harsh exposure of bright, city lights. I actively work to make places darker, but by no means dimmer. By turning out the lights we allow these places to shine. They are a refuge: a safe place for the stargazers, deep thinkers, wildlife watchers, storytellers. To look after the darkness is a privilege. It's the place from which we are all born and to which we will return. I'm a caretaker of the home of our creation and the starlight that exists in us all.

Somehow, as someone who started their life petrified by the dark, born under one of the most light-polluted skies in the UK, I am now a guardian of some of the Darkest Skies currently remaining on our planet.

I have found my place under the starlight and found comfort in our Dark Skies, and that's what I want for you too.

Our lives have always depended on darkness, but it has long occupied a place of fear in human minds. The unknown: a place of secrets and of shadows, hauntings and ghouls. As soon as humans can think, they can also fear. It's a primitive instinct, one that harks back literal millennia, and that for all this time has done an excellent job of making sure our DNA has survived. But why do we still fear the dark?

Humans discovered the first light two million years ago, in Africa, from a flame. Since that first spark, light has revolutionized our lives. Through fire, we not only gained heat, but we also gained light during the dark hours of the day. This was the beginning of our human obsession with banishing the darkness. The heat from the first kept us warm as ice marched across the lands but also gave us sight; glimmers of light during the hours of darkness, offering us protection from teeth- and claw-laden predators, now long extinct. It has vanquished the monsters of the long, cold nights, and the nightmares hiding under our beds. Street lights beam cones of clarity; sentinels selling a concept of safety, standing to attention all night, every night, around the globe, watching the people come and go.

Light has conquered darkness. Arguably it's what made us human. Controlling fire set us apart from the other animals, becoming an indicator of our species' intelligence. I can't imagine the sense of security a fire would bring when we as a species were so exposed, so vulnerable to the elements, to the natural world. It must certainly have made us feel protected and powerful.

This fear of the dark has stayed with us through the ages; through extinctions, uprisings, revolutions and wars. The 'dark' remains, but the threats have changed. We tell our children not to go out after dark, not alone. And if you're a woman, how often were you told not to go out into the night, not alone, not ever?

A violent act on the streets of a city will see a call for action, a call for change. 'We need more lights' is the response our elected officials will reply with (for example, this was the response to Sarah Everard's murder). Because, supposedly, lights equal safety? Not exactly. Additional lights are a visible response. They can say that they've achieved something, that they care about safety, they care about us. However, the danger in the darkness now is not from a wild animal, which would bolt at the strange sight of a human brandishing a source of light. It is crime and it is violence. A moral darkness, untouched by even the most powerful artificial light. All the street lights in the world cannot solve societal issues.

As a young child, I was scared of the dark and struggled to sleep. My bedtime companions were Sleepy Ted (a bear with eternally sleepy eyes), and my glow-worm. The glow-worm was a weird, human-insect hybrid plush, whose plastic face lit up when you pushed its tummy. It had a soft, blue body, and a Wee Willie Winky-esque nightcap topped with a smiling star. It was given to me to 'keep me safe' at night. *Safe from what?* Whatever nightmares small children's imaginations come up with? The monster under the bed? I don't remember what initially made me scared of the dark. What makes our brains conjure up creatures that are only banished by light? It goes back to our primordial years, when darkness could be hiding something that would eat us. Even now, I can't sleep with a foot exposed, dangling temptingly off the bed because, you know, monsters.

We have all experienced lying in bed at night, lights out, only to hear a noise downstairs. You suddenly enter a state of almost painful hyper awareness, your ears straining to detect the slightest noise. Adrenaline courses through your veins as you fail to calm the internal noise of your heartbeat as it pounds against your chest. You wonder how the hell the noise of your breathing suddenly registers at 112 decibels as you try to smother the sounds under the blanket to better listen out for any spurious noises in your environment. You envisage a murderer, creeping up the stairs and intrinsically locate your nearest weapon or hiding place; a stiletto heel perhaps? Or, could I hide in the wardrobe? This fight or flight mechanism is truly hardwired and difficult to control. The mind runs like the clappers and invents 101 things that could be waiting to maim you in increasingly creative ways. You bravely make the decision to spring from your bed, cursing the creaking floorboards as you clutch the stiletto to your chest. Slowly, you creep down the stairs, holding your breath and praying to any god that will listen as you fumble to turn a light on. The grand reveal declares nothing to fear, until you spot the glistening eyes of the cat who's come home through an open window, to gift you with an elephantine mouse – still living – that now resides in your living room.

However, it's not just children who fear the dark. A 2012 study by the British Psychological Society[1] found that forty per cent of adults who participated were too scared to walk around their own homes at night with the lights off. Nyctophobia is the name given to fearing the dark.

My sister, Samara, was petrified of the dark to the point all three of us sisters had to sleep through the night with not only the landing light on, but the bedroom light on, too. I had to have makeshift curtains around my top bunk bed to keep a fraction of the light out. It did nothing to appease my inner night owl, and I would often lie awake, panicking watching the sun come up, knowing I'd have to be up for school soon. Samara's fear, her terror, was very real and very painful for her. The initial switching off of the light, plunging her into an uncontrolled darkness was too much. Her pulse would quicken the minute the switch flicked, panic would clutch at her mind, crawling over her chest to throttle her cries for the light to be turned back on. I don't think there was one, specific thing, but the mystery of not knowing what was out in the darkness is often scarier than the darkness itself.

I have to admit, though I, too, was scared of the dark, I was much less sympathetic at the time (and a lot more tired), and would sometimes turn the light off and remove the bulb. Or throw things at her from the great height of my bunk bed. What definitely didn't help appease her anxieties was when I woke up once in the middle of the night, screaming, because my huge Harry Potter movie poster (*The Chamber of Secrets*, to be precise) had decided to free itself from its Blu-Tack restraints and dramatically envelope my face in high-quality paper. I was nose to nose with Dobby (who was indeed, now free). If you have siblings, or children, you know that one screaming child wakes up all the other children, who also start screaming, which will lead to one or both parents springing haphazardly into action, ready to fight whatever was presumably murdering their offspring, only to find their eldest embarrassing the family name by failing to fight off a paper assailant.

I have had to learn to love the darkness, fighting the urge to be scared of the night. I spent the first seven years of my life on a council estate in Greater Manchester. My mum was often scared half to death as I'd come home from playing out with war wounds from broken glass and other scrapes. Our small garden wasn't safe, either: the fencing panels often being nicked or set alight. During the summer, I'd be sent to bed before it got dark and I'd lie there, fuming at the injustice of it all, as the older kids carried on playing out, often in my own garden. Once, I was so enraged, I tried to climb out of my bed onto the window ledge to see what festivities I was missing, but slipped and fractured my eye socket on the radiator. I looked like I had gone a few rounds in a boxing ring, a giant livid lump on the side of my face and two black eyes. One way to 'see stars', I suppose.

But the estate was a whole lot scarier after sundown. I'd lie in bed listening to sirens and scuffles as fights erupted, slurred arguments and profanities echoing along the narrow brick ginnels. Bottles and pint glasses shattering off kerbs, covering the pavements in a shower of stardust that twinkled under the sodium lights. Closing time at the pub across the road always gave me a knot of fear in my stomach. I'd screw my eyes shut and pray that they would move on quickly, that no one would get hurt. The fear was always worse if I knew my dad was one of the customers the pub landlord was chucking out, and I'd strain my ears, listening, waiting to hear his voice mixed in with the rabble. I'd worry about Tigger, my pet cat, who liked to prowl the streets at night like a small, urban tiger, often coming home injured, his collar wrapped around his body in a figure of eight, contorting his stomach and leaving him unable to use his front leg properly. I was fearful of the things that happen outdoors at night, out of sight. Skeletal fingers stretched outside my window, reaching through the curtains and painting long drawn shadows across the walls. The wind whipping them into motion, bony knuckles rapping the single-paned glass. I would fight with my brain and the panic, to tell myself it was just a tree, not a monster at the window.

When our electricity ran out, the house would be plunged into sudden darkness, and I'd feel panic as the silence rung, somehow hurting my ears. I'd hear a click as the emergency button on the meter was pushed and a *Shit* as the electric refused to come back on. Another click would be the switch on my dad's heavy-duty torch, its beam sweeping across the walls of the hall, like a lighthouse sweeping the sea. Darkness existed in everyday things, too, of course. Sitting on the bottom step of the stairs as kids, we would always do a quick check for welly-dwelling spiders before pushing our feet into their dark depths.

When our leccy ran out, I would always accompany my dad on walks to Trev's corner shop for electric tokens. On the way home, it was my job to hold the small, silver tokens tightly, but not so tightly that I'd cause them to crease and become useless. For reasons unknown, they were made terribly delicate, the slightest scratch on the surface would mean the house wouldn't see light until dawn. But I loved those night-time walks to the shop with my dad. Mainly because it was past my bedtime, and here I was, outdoors, like a grown-up, but also because it made me see darkness in a different, better, way.

Outside, the night changed the world I saw in the daytime. The old, sodium street lamps gave the world a weird, pallid ambience. There was less noise as people went home to bed, with any remaining sounds being more pronounced: footsteps on the pavements, the satisfying thud of a front door being closed, the clink of the chain behind it. The air was crisper, and cleaner, too, as the city's engines were switched off until the morning.

The smell of the night is indescribable. To me, it smells like adventure and rebellion, probably from those memories of being out after dark with my dad. On a really quiet evening, it felt like it was just me and him in the whole world – until a TV would flicker into life through a net-curtained window, and the famous *Coronation Street* cat could be seen making its way along the terraced wall. Another house would be gently lit by the colours of a different channel, and a small table lamp, giving passers-by a glimpse into a different life within each set of four walls. Pictures displaying unknown family faces, China ornaments and floral wallpaper all softly aglow, gave hints to the lives of the people, strangers, who resided there. I was always told off for looking a bit too long.

My favourite thing was finding urban hedgehogs. These nocturnal creatures would often become stuck trying to clamber up kerbs, so we would pick them up – careful not to harm their spikes as they curled themselves up defensively at the gentle, caring touch of a human hand – before finding a new home for them in a soft pile of freshly fallen leaves, sending them shuffling on their way to explore a new home.

It was also on those walks with Dad that he'd often stop and point out the odd star, barely visible through the glow of the great city of Manchester beneath the sky. If I was very lucky, I would get a pack of Fruit Pastilles, with their jewel-like tones of purple and green, and crystallized-sugar coating (maybe this is the reason I never slept), and we would make a detour to the park for a go on the swings. I'd be pushed up into the inky skies towards the glowing moon, that never came closer, looking out across the bright and ever-growing lights of Manchester.

One spring, I remember a great excitement around the estate and lots of people peering out of their windows through net curtains and switching their lights off, plunging the household into darkness to try and gain a better view. My dad took us out on the small green in front of our terraced row and pointed to the night sky above the street lights. At first it was hard to make out, beyond the amber glow, but then I saw it. Or did I? Was it really just that little smudge? Dad walked us out miles, to the edges of the town, where the streets begin to merge into the foot of the Pennines. Now I was sure I could just about see it, clearer but not clear, hanging there in the sky. It looked like a full stop on a blackboard that had smudged, its chalky tail smeared out behind it. This was Comet Hale-Bopp, a visitor from the depths of space, just visible above the rooftops and the glow of the sprawling city.

Very fortunately, my parents managed to scrape the funds together to move us to Wales – Cymru; not just any part of Cymru, but proper heartlands, barely populated Cymru. My very first memory of my adopted home is of darkness, when Uncle David drove us there at night in a hired truck along the A55, I stayed awake long enough to see the Marble Church at Bodelwyddan. A beacon on the roadside, its beautiful 202-foot-high white spire illuminated by huge flood-lights below it, the church is seen for miles before you reach it. The light bleaches it of all detail and the marble glows, almost glaring. I fell asleep, only waking when the truck came to a head-jostling stop. Bleary-eyed, I looked around and thought we must have pulled over for some reason. All I remember is intense darkness. Surely this couldn't be where we were moving to? There was nothing to move to! Walking up to the front door, everything felt so silent and still. No lights welcomed us through the window, and no one had put money on the meter, so I wandered blindly around the inside, scared to put one foot in front of the other. The air smelt so clean and fresh it stung my nose and hurt my lungs and it was so *quiet.* I could faintly hear a roaring sound that seven-year-old me didn't recognize as the sea; the booms as waves crashed onto the shore sounded like the fizz of huge fireworks. I then walked straight into a glass door, thinking I was stepping into open space, which also stung my nose. My grand entrance to Cymru was a squashed nose and darkness.

That night, we lay under blankets on mattresses on the floor, too excited to sleep. I pressed my face up against the window and saw nothing but deepest black, reaching out for eternity, until a thin tower of red lights was visible in the distance. My dad was excited to tell me that it was a mast at Nebo, just outside of Snowdonia National Park and the place where his uncles had been evacuated to during the Second World War. I had no idea what a national park was or meant, but one thing was for sure: I had truly arrived in darkest, deepest Wales, and in the best of ways.

From then on, my life changed, forever. I left the house at breakfast and got home when the street lights came on. I remember running at what felt like break-neck speed as they sprang into life, one after the other behind me. The threatening 'ping' as the elements heated and burst into life, ready for another night shift, filled me with fear as I knew I'd be in trouble if I didn't beat them back.

I made friends at the local school, and we spent our time swimming in the sea, running through fields or sneaking into the posh gardens of empty holiday homes (we knew exactly which ones had trampolines). Wading through Afon Crigyll, damp trousers rolled up above our knees, I first recoiled as huge brown fish rolled around just under the surface – the rainbow-coloured backs never quite breaking through the watery divide between their world and the sky. Flat fish would squirm and wriggle underfoot, and we'd squeal at the shock and the tickle as they shot off, letting a sandy trail settle in their wake. It took a while to settle into this nature-filled place, so used to being hemmed in by dusty roads and grid-locked streets that we didn't understand how there was so much space to grow into, to roam into. It was unsettling at first, like having the strict rules removed, revealing a world we didn't know how to navigate, without pavements from one point to the next. This extended into the night skies full of endless stars, a whole new universe to explore after sunset. Lying in the long grasses under shooting stars, refusing to acknowledge the cold seeping up from the ground as we longed to see just one more meteor streak through the dark night.

I was a world away from the dangers of the city and I relished my new-found freedom, but I was never as wild and carefree as my new set of friends, who seemed truly fearless, swinging from lamp posts and scaling heights that would have made their mothers faint had they known. I couldn't even ride a bike, yet they rode around with no hands, eating sweets and giving backies all at the same time. But when the nights came, even these brave warriors retreated home to safety. It seemed in the end that my new-found fearless friends were still afraid of the dark.

My dad had always been interested in stars, but now that we lived in a naturally dark place I was dragged out for walks on every clear night; albeit walks with Dad usually meant watching him disappearing over the horizon while I'd be lagging behind, feet aching, fingers freezing and tears streaming from the cold. I'd be bundled up in coats and scarves and waddled down to the beach where we would sit for hours, letting our eyes slowly adjust as they gathered more and more starlight until the Milky Way streamed above us. There would be no sounds from people. Just the sound of waves, crashing onto the shore, and seabirds calling from their islands in the sea. The wind would whip my hair around my face, turning my nose and cheeks a rosy red and teasing the boats to the shore, their ropes making a gentle *tink tink tink* against the metal masts. Summer nights could be balmy here; in the stillness, not a single ripple on the surface of the sea.

We replaced Trev's shop with Pete's on our night-time walks, and I would get a cracking penny mix and we would walk to an abandoned house sat out on a little bay, its weathered window frames holding nothing but the views of the sea and the stars, and sit on its crumbling front step, its concrete radiating a little warmth it had collected from the sun that day, and Dad and I would sit and stare. Sometimes we'd see lightning lash out at the Irish coast, as warm and cold air collided. Tridents forked in purple, blue and green, piercing the sky. No rumble of thunder would follow behind, just flashing lights out at sea. I'd never known peace like it. It was a far cry from the sirens and violence of our old estate in the city.

It was the heady days of the beginnings of the internet for the masses. Dad would go to the local library and spend hours painfully loading up pages on the leisurely dial-up connection. Line by line the screen would slowly reveal what to see in the night sky, while I spent that time loading up pictures of Pokémon to print out for absolutely no reason at all. At five pence a sheet, I worked out I'd need £7.55 to complete my paperback Pokedex, and on the dial-up internet, probably 755 years. (It was never completed, I didn't even reach Jigglypuff.) Dad would then tell me what we could see, the myths and stories of the constellations as Orion appeared in winter and the Great Bear in spring. On BBC Radio 4, Sir Patrick Moore's distinct voice would travel down the airwaves and talk about space in a way I didn't really understand, but I listened anyway. Dad would take notes, alongside the print-out sheets. These fast became my favourite times. As one of five children, it was a rare opportunity to get some time alone with Dad. Having my tiny mind boggled by the size of the universe and realizing how insignificant I truly was, was weirdly terrifying and comforting all at once.

As my confidence in the dark grew, I was also amazed at how much I didn't need a torch. Our way was lit by stars and the moon. Some nights the moon would shine so brilliantly it created a silvery path all the way out to sea. Dad, always full of it, would tell me he could see where man had landed on the moon in 1969, but this was when my love and appreciation of Dark Skies started. I will always be grateful that he passed on his knowledge and that he never discouraged me from going out at night. 'You're the scariest thing out there,' is what he'd tell me, when I was forever trying to keep up with his giant strides as he disappeared into the night.

It has definitely taken me time to overcome a fear of the dark, and still I can't fully control my inner Neanderthal when it comes to an unexpected sound from a cow I don't realize is stood six feet away from me in the darkness, or I see a weird inexplicable shadowy movement in the corner of my eye. That same panic takes grip, my chest tightens and adrenaline runs rampant, but now I can talk it down, and put my prehistoric self back in her cave. Most of the time at least.

⋙◎⋘

In the UK, we should all be a lot more appreciative of the darkness we have; living so far north and with such long winters, we are blessed with extended nights and many hours of darkness.

Come December it would be dark as I stepped off the school bus at quarter past four. Collectively, we children counted down the final days of summer, mourning its loss as the clocks fell back, and the nights leaped ever forward, darkness gaining at what felt like an hour a day as we hurtled into the depths of winter. Now I know that we should celebrate that time where the hours of dark outweigh the light, like the ancient Britons did all those years ago. We should welcome that restorative night, allow it to calm the hot heat of summer, let it balance us back into a reflective state. Time to slow down into that serene shelter of solstice and unwind, even semi-hibernate if we want to.

Rather than spend half the year in mourning for the summer past and waiting for the summer future, we should embrace our winters; embrace the darkness, get outdoors into our gardens and parks and onto our streets, to learn what the night has to offer. With ninety-nine per cent of the Western world living under light-polluted skies, it's no surprise we have forgotten we should have front-row tickets to the Theatre of the Night, the greatest show not-quite-on Earth. Let's save our Dark Skies and bring the stars back into focus, for everyone, everywhere.

Our lives depend on it.

CHAPTER TWO

This Little Light of Mine

There are two kinds of light – the glow that illuminates, and the glare that obscures.

– James Thurber

The Glover Review, as it's come to be known, was a review written in 2019 of the National Landscapes in England on behalf of the Department for Environment, Food and Rural Affairs (DEFRA), by Julian Glover OBE. National Landscapes are our family of national parks and Areas of Outstanding Natural Beauty (AONBs), and the report was tasked with detailing their current condition and how well they work as breathing spaces and places for the nation to visit. In the days after the review was released, the headlines focused on this snippet, taken from Proposal 8 of the report: 'Every child should spend a night under the stars'. The reasoning being that too many have lost their connection with nature, and what better way to rekindle that connection? What could fire up that urge to protect and care for our natural world more than a night tucked in under a magical blanket of thousands of stars? Half the park is after dark, after all. The mountains don't simply settle into slumber, nor the rivers stop their relentless running from summit to sea, just because the sun sets and the humans disappear. To see our most treasured landscapes in the daylight tells just half their story. After the watershed, when we're all tucked up in bed, a whole new shift begins in our national parks – the night shift. The nightscape should be just as coveted as the landscape itself.

The report dances around the importance of Dark Skies and how they are a vital asset that should be accessible to all. The need to protect darkness is right there, on the tip of his tongue, but frustratingly Glover falls just short of saying that Dark Skies are essential for a healthy nation. They enthral and inspire us. They are humbling to witness. To spend a night under the Milky Way is to open your mind to the knowledge that we are made of stars, and they are made of us. Everything we can see, hold, touch and breathe has come from the stars, the matter of the universe, including your lungs, which sting from that crisp night air and the creatures whose calls soundtrack the night. The report fails to ask the bigger questions. How can every child spend a night under a Dark Sky if darkness is disappearing?

Eighty-three per cent of the world's population is living under light-polluted skies. If we look at just the US and Western Europe, those numbers leap up to ninety-nine per cent of the population with about fourteen per cent of people no longer using their night vision.[1]

In the UK, a staggering ninety-eight per cent of us are living under light-polluted skies. In our biggest cities, people can hope to see around ten to twenty stars, out of the 5,000 that should be visible to the naked eye.

It's a disaster for darkness.* Light pollution has increased by at least forty-nine per cent over just twenty-five years, with some scientific estimates putting the global increase at a ginormous 270 per cent.[2] This discrepancy in reporting is due to the satellites that we use for monitoring light-pollution data being 'blind' to certain light wavelengths and sources; for example, digital billboards and windows, which will impact the view to the stars of a person standing on the ground. These figures are set to increase even further, with light-pollution expert, Christopher Kyba, finding that 'the average brightness of the night sky is increasing by ten per cent every year'.

Kyba also fears that Orion, a staple of the winter night sky and one of our brightest constellations, will no longer be visible to most of us in the very near future if we don't

* The etymology of 'disaster' itself meaning 'ill-starred' or 'lack of stars'.

halt the current trend of exponentially rising light pollution. The unthinkable, a starless sky.

If you've never heard of light pollution, you're not alone. Light pollution is one of the least understood pollutants there is, with a small cohort of people working hard to bring it to the attention of the masses and demand action to reverse our light-hungry ways.

In the UK, there is work ongoing to protect the surviving areas of darkness, such as the Dark Sky Reserves of Eryri (Snowdonia) and Bannau Brycheiniog (Brecon Beacons) National Parks in Cymru. In the whole of the British Isles, just twenty-one places have protections in place and are accredited as being part of the International Dark Sky Association's Dark Sky Places programme. They most prevalently reside in our national landscapes, where planning controls on lighting can be implemented at a local level.

Our national parks and AONBs are supposed to be oases of tranquillity, where the nation can spend time in the raw elements of nature – darkness included. When we talk of protected landscapes, we use terminology like 'character areas' and 'visual impact', yet so many are slow to realize that 'half the park is after dark'. Artificial light at night has a massively negative impact on the character of our most beloved landscapes, transforming them from some of the only places left on Earth where you can enjoy natural darkness, uninterrupted by glare and sky glow, into fragmented and damaged night-time environments, split apart by pools of light. In the rural setting, just one powerful flood-light can ruin a nightscape, visible to all from miles around, a constant reminder of the human footprint.

Some of our protected landscapes are already under threat, like the Peak District National Park, whose hills have watched on as cities – Manchester, Sheffield and Leeds – have sprung into being at its feet, powering the industrial revolution from the factories nestled in its arms. But from the tops of Kinder Scout, where the feet of many once trespassed for our right to access the great outdoors, the only trespasser now is light. A trespasser who should rightfully be sent packing. This hasn't stopped the national park from tackling the issue, protecting the dark pockets that remain and talking to their neighbours to explain how they're impacting not just their own areas, but one of the most important and historic national parks in Britain, denying huge swathes of our population the right to an unpolluted night sky. If the night sky above the Peak District could be restored, that would open up the universe to the twenty million people who live within an hour of its national park.[3] That would go some way in realizing Proposal 8.

Cymru has the highest percentage of protected Dark Skies of anywhere in the world,[†] with a constellation of

[†] 17.6 per cent of our skies so far.

reserves, parks, and Europe's first Dark Sky Sanctuary on Ynys Enlli (Bardsey Island). (Yes, they say that Wales is stuck in the dark ages, but we love our darkness, thank you very much.) This is partly thanks to having two sizeable national parks right at our core, and helped further by the fact we have retained wide areas of rural land, untarnished by development. We are not ones for resting on our laurels, and here we are continuing to push for national laws and controls on light pollution that will benefit our residents in all twenty-two unitary authorities, creating a healthier, happier society; one that allows each of its residents to access what should be accessible to all – the night sky. In turn, it will save the country financially, in carbon savings; and protect our biodiversity, returning the night to its natural state and handing it back to the wildlife who depend on it.

Sadly, light pollution continues to threaten even our protected places, as the creep of light pollution grows from the areas at their boundaries. There's only so much that can be done within their boundaries to control lighting, but if our neighbours continue to be inconsiderate and careless with light use, the dark hearts of our islands will soon fall, too. Currently, half of England's very darkest skies are vulnerable to erosion by light pollution, with no legal protections at all.[4] Some areas have not acted as they don't qualify for the International Dark Sky Association's programme, but no area should be striving to protect darkness for an accolade alone. We should protect darkness because it is the *right* thing to do.

There is next to no work being completed on implementing light-pollution controls in our urban areas, yet these populations and communities are the most vulnerable to the health impacts of light pollution. They are already the most likely to be struggling to access wild spaces and green places; and for them, even access to space itself has been blocked. While we may not be physically able to barricade access to the night sky, like we can open fields and hillsides, light pollution has stolen the night right from under the noses of most of our population while it sleeps. More shockingly, in many areas, they are already the second or third generation for whom the stars have been placed beyond reach. The starlit skies are locked away in the memories of their grandparents, seen only on the pages of bedtime stories or TV screens.

Yet there is little to no appetite for change from most local authorities and elected officials overseeing these most polluted places, despite an outcry from their residents worried about the health of their communities. Lighting has fallen victim to the tick-box exercise of cost and carbon savings, and that's the bottom line for most councils. You can't care about what you can't see, and they are certainly not seeing stars. At some point, the government will awaken to the needless damage that has taken place on the streets they represent, but so much of this damage won't be able to be undone unless they act fast and act now.

We have yet to fully understand what this remarkable change to our environment and lifestyle will mean for our biology, but it's already known it has dismal consequences for human health.

How did we get here? And how has it happened so fast? First, we need to know a little history about the humble light bulb.

Without light and dark we would have no night and day, no sunsets, no sunrises, no natural cues to let us know the time of day. For eons, we mere mortals were tied to the whims of the sun. It dictated when we got up and on with our lives and when it was time to settle in for the evening, and that was the way of things for a long, long time.

Then along came Thomas Edison. Deaf in one ear and deemed 'difficult' by his teachers, he attended school for a grand total of twelve weeks. His mother, Nancy Edison, took on the task of educating her son. An accomplished teacher, Nancy had a massive influence on Edison's life, and it is without doubt she who equipped him with his ingenuity. Despite a less than glowing academic record, he is deemed one of the greatest inventors of history, revolutionizing the way people were living, and influencing how we live today. He held over 1,000 patents for his inventions, including the phonograph and, of course, the light bulb.

The story of the modern light bulb is similar to the Goldilocks fairy-tale. There was the English inventor, Sir Humphrey Davy, but his work was too expensive to be viable. Then along came another British inventor, Warren De la Rue, but his bulb burned out too rapidly. There are dozens of inventors who have all helped the light bulb on its way to our homes over hundreds of years, but they were all too big, too expensive, or too flammable. Finally, in 1879, Chemist Joseph Swan successfully demonstrated a working incandescent light at a lecture in Newcastle, England.

Before electric lights, the streets were lined with gas lamps. Gas lanterns have become iconic and synonymous with Victorian Britain thanks to films like *Mary Poppins* and *Peter Pan*. They were beautiful and stylish, elegant steel structures, with gentle flickering orange lights. However beautiful they were to look at, they were unfortunately quite deadly to live with, giving off terrible fumes and blackening walls when used inside homes. There was also the added gamble that they may surprise you with an explosion, not really conducive to a relaxing home atmosphere. In the early 1800s, it was a common sight in British towns to see lamplighters. They would walk the street with a ladder and a hand lamp, setting the streets a glow. They had quite the job to do; by 1823, 40,000 gas lamps covered 200 miles of streets in London alone. The gentle light of the gas lamps was just enough to see by and gave British cities that Dickensian charm that so many tourists are disappointed to see has disappeared, along with the bowler hats and crinoline skirts.

Briefly, we had the terrifying 'electric arc' street lamps. They got their name due to the way they worked – an electrical current would arc between two carbon rods, giving out a hugely intense, but unstable, light. They were so bright and burned so hot that they were installed high above the streets, some on poles of near to 150 feet. They created a loud, buzzing noise, often sending sparks to the ground, causing injuries and fires. Compared to gas lamps, they were phenomenally bright. Gas lamps measured at around twelve candles[5] (roughly 156 lumens)[‡], whereas the monstrous arc lamps weighed in anywhere between 10,000–100,000 candles (120,000–1.2 million lumens)[§]. That's essentially enough light to give yourself an X-ray, or look into the future. They were so strong; women used their umbrellas to protect themselves from this artificial sun.

These lights were, unsurprisingly, unpopular. Robert Louis Stevenson, the author of *The Strange Case of Dr Jekyll and Mr Hyde*, was not a fan.

[‡] For comparison, a modern LED light bulb for household use is 13 Watts with an output of 1521 lumens (with a 200 degree rated beam angle).

[§] For the lighting engineers reading this, this is a rough conversion of candles to lumens and does not take into account any directional component which would alter the calculation.

> *A new sort of urban star now shines out nightly, horrible, unearthly, obnoxious to the human eye: a lamp for a nightmare. Such a light as this should shine only for murders and public crime, or along the corridors of lunatic asylums, a horror to heighten horror.*

In fact, it seems that Stevenson was something of a defender of Dark Skies in his time. He wrote often about lighting, with an essay 'A Plea for Gas Lamps', and the poem, *The Lamplighter*, suggesting he was much fonder of this nostalgic street lighting.

Still, the quest to find a commercially viable and nonlethal lighting solution was burning bright. It was Mosley Street, Newcastle, that became the very first street to be lit by incandescent electric light bulbs. On 3 February 1879, this Georgian street was forever changed with the flick of a switch. Unfortunately for Swan, his lamp was also impractical for commercial use. Doubly unfortunately, Edison figured out where Swan had gone wrong and so it was Edison who won the race and patented the first commercially successful incandescent light bulb. To be fair to Swan, Edison had research staff to whom he could hand over the duties and, during a couple of years, tested over 3,000 different designs for bulbs and over 6,000 different plant types for use in the bulbs' filaments. I'd say that's quite a competitive edge.

Swan and Edison (after a legal battle), eventually joined forces to create Edison-Swan United, becoming the world's largest manufacturer of light bulbs. Edison demonstrated his bulb on New Year's Eve 1879 and thus, my arch nemesis was given life, 110 years before my birth.

The light bulb was one of the most significant advancements since humans discovered fire, revolutionizing society, and our way of life in ways that we now take for granted. No longer ruled by daylight hours, we could stay up long after sundown. It was so revolutionary, that when a new shopping parade opened in Brixton, London, in 1880, resplendent with electric street lights, people wandered in wonder under its glass canopy of an evening. The street was even named Electric Avenue, to mark this marvel of modern technology, now of course synonymous with the Eddy Grant song of the same name. I can imagine now, the ladies and gentlemen of London dressed in their best dresses and tailcoats, walking canes tapping on the artificially bright street surface. It must have felt like actual magic.

Behind every modern advancement for perceived societal good, there are usually more profit-focused drivers behind the wheels of change. Electric lighting was no different. For the average working person, the (usually brutal) workday came to a close at sunset. Most factories were dark and dangerous places, dimly lit by gas lamps that needed constant attention, let off noxious smells, and of course had the added peril of burning places to the ground, which wasn't good for profits. It was often not financially viable to keep workhouses open past dark. This meant the working classes had quite a bit of leisure time, to rest and recover from their hard days labouring, albeit in spaces dimly lit by candlelight. This all changed with the advent of the light bulb. With pound signs in their eyes, factory owners immediately saw the advantages this could bring their businesses. They were quick to install this new lighting, delighted that work shifts could continue all through the night. This was great news for factory owners, whose productivity and profits soared, lighting the way of an already booming industrial revolution. It was not such great news for the working classes, for whom these new-fangled light contraptions meant that cities never slept again, inspiring the idea that sleep was for the weak. Demand for workers spiked as factories doubled their outputs. In the twenty years following Edison and Swan's light bulb, New York City's population ballooned by more than 1.5 million, as people left the rural areas to find work in the newly lit 24/7 city. Slowly, the gas lamps and candles were replaced at work, and much later in the home. At least it was a safer environment that was less likely to turn to ash.

People fear change, and they certainly feared this new 'electric sun'. Little did they know, their concerns about this unnatural light source and its impact on our natural flow of life were very nearly spot on, but it would take close to a century before we started to understand the hidden damage that all this light has done. One of these implications revealed itself a little quicker than the rest. It is some coincidence that Thomas Edison, himself, hated sleep. He hated it so much that his bulbs of brightness have inadvertently wreaked havoc on the circadian rhythm of not just humans, but our wildlife too. The circadian rhythm is our internal metre that tells us when we need to sleep and when it's time to rise and shine. Edison believed that sleep was an inherent and repulsive value in human beings who were just lazy good for nothings:

> *People will not only do what they like to do – they overdo it 100 per cent. Most people overeat 100 per cent, and oversleep 100 per cent, because they like it. That extra 100 per cent makes them unhealthy and inefficient. The person who sleeps eight or ten hours a night is never fully asleep and never fully awake – they have only different degrees of doze through the twenty-four hours . . . For myself I never found need of more than four or five hours' sleep in the twenty-four. I never dream. It's real sleep. When by chance I have taken more I wake dull and indolent. We are always hearing people talk about 'loss of sleep' as a calamity. They better call it loss of time, vitality and opportunities. Just to satisfy my curiosity I have gone through files of the British Medical Journal*

and could not find a single case reported of anybody being hurt by loss of sleep. Insomnia is different entirely – but some people think they have insomnia if they can sleep only ten hours every night.

Edison claimed to only need an upsettingly meagre four hours of sleep a night. With the invention of the electric light, humans believed they had solved the need for sleep. Pre-light bulb, humans slept an average of ten hours a night, a healthy reaction to the natural, lighting cues of their world. Today, the average human gets around seven hours sleep, with one third of Americans not even getting six hours of sleep a night.[6] As life has become about 24/7 hustle culture, we no longer value sleep as a society. I don't know about you, but I'd be positively murderous on Edison's four hours, but clearly it was the fuel to his spectacular brain power. Or was it?

We now know sleep is vital to us if we want to function properly. Deep down, Edison knew this, too. For all his boasting about not needing sleep, he was often found (and photographed) taking 'power naps' in various places around his labs and in specially placed cots for when his brain needed charging. Sleep is crucial to our mental and physical health and wellbeing. For babies, children and teens, sleep is vital to ensure they develop and grow properly. But more about the science of sleep later.

Since that first street in Newcastle, we have disrupted the natural cycles. We have hoisted street lights everywhere. Roads, motorways, alleyways, parks, beachfronts and waterways, castles and ruins, all are embellished with a light bulb, or ten. In our pursuit to defeat darkness, we have led a relentless tirade, rallying against the night. We have strung the British Isles in festoons of lights; our road networks, when seen from above, glitter like a golden spider's web. But all this light has led to an issue no one saw looming in the shadows: light pollution. But what *is* it?

Light pollution happens when useful light that we pay good money for and use good energy to make, escapes from badly designed and placed light fittings. Instead of the light being focused on where we need it and on what we want to see, this light is wastefully sent skywards. It's easily noticeable on cloudy nights, as this wasted light illuminates their bases, reflecting it back towards the ground. You'll see clouds glowing orange and white as they float around our towns and cities at night. This waste of light is known as skyglow or 'Artificial Light At Night' (ALAN). If I stand up above the dramatic cliffs of Ynys Lawd, Caergybi (Holyhead), upon which the famous lighthouse is perched, looking out across the Celtic Sea, I can see, distinctly, an orange dome of light, enclosing Dublin (Baile Atha Cliath) like a giant cloche surrounding the city. Impressive, considering Dublin is a good sixty miles away.

What use is all that light escaping into the sky? It's of no use at all. In fact, it is nothing more than a waste of money and energy.

But Dublin is not alone in its wasteful ways. Far from it. In fact, over eighty-five per cent of EU territory is affected by artificial light.[7] Your mam was right. ('It's like the Blackpool Illuminations in here!') If you care to view the world via a light-pollution map, you will see just how bad the issue is. Mainland Europe is lit by over sixty million street lights, which amounts to one per person in the UK. On a night map, the forms of many cities and towns are outlined by tinselled borders of light. We create all this light, yet we have frustratingly little control over it, and that's the crux of the issue.

But there is hope. The Nightlight Project combines several EU countries, who are working together to create policies to reverse light pollution in Europe. In France, they deployed an impressive lighting curfew to curtail the harmful impacts, with all non-essential lighting being switched off at midnight. This means billboard, supermarket, and shop lighting (basically anything you don't need to find at night) is switched off. All essential buildings such as hospitals are of course left lit, but this simple curfew in its first year saved French Authorities 250,000 tons of carbon and £166 million in energy costs.[8] Phenomenal. To offset that carbon, you'd need to get busy planting trees – around 1.2

million broadleaf specimens a year, in fact. This is exactly the sort of action we need. If it can be done in France, we can do it here too.

In Britain, we are awash with light from around seven million street lights. London alone is home to well over 500,000 bulbs, not including billboards, skyscrapers, or other domestic lighting. That's a lot of light for such a small island and the reason why only two per cent of the UK population receives a truly Dark Sky.[9]

I moved from Ynys Môn (Anglesey) to Trefforest in 2011 to attend the University of Glamorgan.[¶] My halls were in a semi-urban location, high on the hillside of the valley with views up towards Pontypridd and down the A470 towards Cardiff. This was my first time away from home and a big move. I was homesick and incredibly overwhelmed. It's lucky I couldn't afford the ticket, or I'd have been straight back on the train to the island. I went outside to look at the stars, something I will always do when things seem to be spiralling. I walked a way up the hill above campus, cutting through a broken fence into the trees, and looked up. To my horror there were no stars. I couldn't see a single one. The light pollution was so severe, the campus was submerged under a pallid, yellow glow. I noticed the flood-lights that

[¶] Now the University of South Wales.

blasted all darkness into submission dotted around the paths and car parks, figures of people coming and going around the buildings, the powerful white beams washing out any descriptive features. I felt very unsettled that my window to the stars was blocked, my one constant was gone. From the A470 the Trefforest campus is immediately recognisable for its white domed top, an observatory.

One evening I got talking to a group of astronomy students, a course I was unaware the university ran. I was excited to ask them about the observatory but was saddened to learn they were the last cohort of astronomy students at the university. The course would be discontinued after their graduation, light pollution and an observatory in need of much maintenance, two of the cited reasons for ending it. That was the first time I heard the term 'light pollution' used. I now had a name for my foe.

The UK has been rapidly changing all our street lights to bright white LEDs (Light Emitting Diodes). The LED frenzy began when local authorities quite rightly started to look at ways to reduce their carbon emissions and lighting costs. When they realized just how much money could be saved by switching from the old sodium lamps, the roll out began. However, as good intentioned as the local authorities were, they didn't realize the negative implications of the switch until it was too late.

To be fair to councils, LEDs have made great savings financially as well as to emissions, for example, look at the figures from Northumberland County Council:

> *Northumberland County Council has almost converted all its 46,500 lights to LED, which, combined with dimming lights, will save 6,675 tons over the lifetime of the project. It is reducing energy consumption by 63.6 per cent, which equates to a current annual saving of £1.25m.*[10]

You can't argue that these aren't astronomical savings, but doesn't it seem too good to be true? Sadly, it is. When the switchover started, there was little knowledge in the public realm about light pollution, and even less so about the damaging effects of the bright, white lights most councils opted for, which measure around 5,000–6,000 Kelvin on the colour temperature scale.

What's the issue with LEDs and what is a Kelvin, you ask? The blue LED was created in the early 1990s by Isamu Akasaki, Hiroshi Amano and Shuji Nakamura, a group of Japanese-American scientists, so it is relatively new to us. In 2014, the trio were awarded the Nobel Prize for Physics for their creation. The blue LED allowed scientists to create white LED light bulbs by coating the diodes with phosphor, giving that sharp, clinical, white glow you see with many street lights being used today.

Unfortunately, we now know that the blue-wave light emitted by bright white LEDs are damaging. During the daytime, they can sometimes be useful because they enhance attention and reaction times and can be a mood-boosting factor. However, there's a 'dark side to blue light'. According to a Harvard medical study:

> *While light of any kind can suppress the secretion of melatonin, blue light at night does so more powerfully. Harvard researchers and their colleagues conducted an experiment comparing the effects of 6.5 hours of exposure to blue light to exposure to green light of comparable brightness. The blue light suppressed melatonin for about twice as long as the green light and shifted circadian rhythms by twice as much (3 hours vs. 1.5 hours).*[11]

A French health agency also recently reported that blue light can cause physical damage to our eyes, with those aged over fifty, and children whose eyes are not yet fully developed, at heightened risk.

> *New scientific evidence confirms the "phototoxic effects" of short-term exposures to high-intensity blue light, as well as an increased risk of age-related macular degeneration after chronic exposure to lower-intensity sources, according to the French Agency for Food, Environmental and Occupational*

Health & Safety, known as ANSES. Age-related macular degeneration, a leading cause of vision loss among people over fifty, causes damage to the macula, a small spot near the centre of the retina that's needed for sharp central vision.[12]

There have been complaints about these new street lights in cities the world over. People find it too clinical, too bright and glaring. In Rome they've even accused LEDs of killing the city's romance and residents in Davis, California have demanded a total refit from this new harsh lighting.

LEDs are up to ninety per cent more energy efficient than the old incandescent bulbs, which is fantastic. They're also incredibly cheap to buy and can last up to four times longer than their filament bulb ancestors. The downside to this, however, means that increasingly high-powered lights are coming to market and being installed in domestic settings, where they are totally inappropriate. Most people will go to their local hardware store and buy the cheapest and most powerful light they can afford. This has led to a plague of overpowered glare monsters being installed up and down the country, blinding neighbours, pedestrians and drivers, and casting swathes of light into what was once, dark wildlife habitat. No matter how cheap a light is to run, if it's creating too much light for the job, you're still being inefficient, still wasting your money. And the cost to our health and wildlife is just too high to ignore.

We now know that it is much safer for human health to be exposed to lower temperatures on the light spectrum, this means less of that pesky blue-wave light hitting your precious retinas. Doctors, lighting professionals and the International Dark Sky Association are calling for lights to be lower than 3,000 Kelvin (the unit used to measure the colour temperature of light) with 2,700 Kelvin recommended as best for our eyes, and even more importantly our wildlife. This doesn't affect the light's brightness, or how much coverage it has, but simply its colour temperature, creating a gentler and much more appealing warm, amber light.

By rushing to reduce carbon emissions, we were blinded by the lights. We forgot to look at the other implications changes to lighting can have on biodiversity and our health. We need to be very careful we don't dig ourselves out of the climate crisis by creating an ecological collapse soon afterwards.

The fittings and mechanisms for lighting also play a part. Dusk till dawn sensors mean lights come on as soon as the sun goes down and burn electricity throughout the night, whether we really need the light to be on or not. The majority, if not all, of UK street lights work on these sensors. We need to ask ourselves, *Are all these lights on throughout the night really useful? How many streets are lit throughout the night for no one to walk on? Can we afford to turn some off?*

My argument is that we can't afford not to. In the UK alone, light pollution costs the UK Government (therefore you, the taxpayer) £1 billion a year in wasted energy.[13] In Europe, it is 7 billion euros annually. Local councils were estimated to spend £613 million on street lighting in 2014–2015, accounting for between fifteen and thirty per cent of a council's carbon emissions.[14] This doesn't even begin to include illuminated signs, billboards, shopfronts, EV charging points – the list goes on.

On a global scale, light at night accounts for fifteen per cent of the global energy supply and emits five per cent of global greenhouse-gas emissions. That's a huge waste of resources when we are facing climate collapse. We need widescale rethinking of the lighting of populated places. According to the UN, simply switching to more energy-efficient lighting, like LEDs (of the appropriate colour temperature), would save enough energy to retire 250 coal power plants.[15]

LEDs used in the right way, and at the right colour temperature, could be a really quick win in the global fight against climate change. However, it's more than just swapping to LEDs. To ensure lights are as efficient as possible, there can be no upward spill of light. Having an energy-efficient bulb is no good if you're still wasting energy and money by letting light spill out where it isn't needed, like into the night sky or onto a neighbour's property. By using a fully shielded light fitting and pointing it downwards, the bulb can no longer create glare. You will have much greater control over the spill of light and be able to illuminate exactly what you need to.

Despite light pollution being so easy to tackle, and despite valiant efforts from various organisations, light pollution is increasing globally at, quite literally, the speed of light.

Where Dark Sky places exist, planners will insist on four simple rules to ensure there is no light pollution impacting on the night-time environment.

1. Lights lower than 3,000 Kelvin.
2. Lights to be fully shielded and downward facing and less than 50W.
3. Motion sensors or timers should be used, rather than dusk until dawn sensors.
4. If it's not needed, switch it off. Light what you need, when you need it, with the right amount of light.

It's that easy.

The current human relationship to light is a complex and currently toxic one. We need a rehabilitation to stop us making the same old mistakes and to realize the true impact of this unhealthy infatuation. There are healthier ways to light our world that don't harm our environment or our health and wellbeing. The solution to light pollution is staggeringly simple, yet we are being painfully slow to make the changes. The benefits for nationwide Dark Sky lighting policies being implemented speak for themselves, and with so much in the balance, we cannot afford inaction. Only when we act as a whole will Proposal 8 of *The Glover Review* be able to be recognized. Until then, most of us are left in the dark about what's really out there in the night. You're being denied your right to a dark and starry sky, go out and fight for it.

And turn that bloody light off, will you?

CHAPTER THREE

Away with the Fairies

Humans spend a third of their lives asleep. That's around twenty-six years of sweet dreams for the average person. For some of us, sleep comes easy. Like my nan. Every Sunday, Uncle David would scroll through football results on Teletext at the rate a snail edges along a garden path. The clock on the wall ticking, getting louder the longer I was sat there in the living room. I'd be itching, incapable of being still in the silence. Boredom was a physical pain, like my brain was going to bulge out through my eyes. My nan would then utter the dreaded words, 'I'm not going to sleep, I'm just getting my forty winks,' and just like that, she would be away, melted into her green velour armchair, catching flies, feet up on the battered leather pouffe I would pretend was a turtle shell. She could do this anywhere; it was an impressive skill. But knowing it would be a good hour until anyone did anything remotely interesting,

I would deflate. I'd resign myself to leaning, arms folded, on the back of the sofa, watching the hedge sparrows play happily in the puddles through the window, the sound of their chitter-chatter chirping against the electrical hum of the TV set, where the pixelated lime-green text wavered on the screen, declaring Oldham Athletic had lost. Again. I did like the owl on the club's badge, though, it's knitted eyes stared out from the blue-and-white scarf slung about my uncle's shoulders.

For others, falling asleep at the drop of a hat is a skill we could literally only dream of. I am in the unfortunate camp of the eternally exhausted. I've never been a good sleeper. Practically from the moment I was born, my poor parents took shifts, day and night, as I just didn't sleep. My mum once sheepishly admitted that when I was six months old, she took me to the GP to ask for help, and they gave me medication. I don't know why she is embarrassed by this. I have told her many times that I wouldn't have judged her or Dad if they had set me free out into the Pennines to live as a feral wolf-child. There cannot be a punishment worse to parents than a child who refuses to sleep. I am told, reliably, that other than that, I was an ideal child. Probably because I'd been sedated. The family albums contain photos of me, wide awake on the lap of a parent who is either asleep or fighting it. Most evenings my dad strapped me into my buggy, pushing me for miles around Greater Manchester, willing me to fall asleep, while I refused to close my eyes. If I'd been him, I would have left the buggy at a police station and got a few hours sleep while they figured out who the child belonged to.

I don't understand why sleep has always evaded me. I have grown to accept I will never be one who arises with the morning chorus of birdsong, full of the joys of spring. I belong to the night. Unfortunately for people like me, society expects us to be functional early in the morning, ready to work by nine o'clock sharp. It seems unfair that if I get up late, I'm considered lazy. Why is it that those who go to bed early are not viewed the same? Why are daylight hours considered more acceptable for activity than twilight and night-time hours? I am still active for the same amount of time in a twenty-four hour period, but because I don't fit the social norm of nine-to-five, society brands me idle. I have tried everything: tablets, herbal remedies, balms, sprays, bath salts, Horlicks, staying awake for forty-eight hours to trick myself into a 'normal cycle'; but nothing works. I'm just not wired that way ...

'Go to the toilets and wipe that make-up off Danielle,' scolded my head of year at school.

'But I'm not wearing any, Miss.'

I was sent out, again, to scrub at my eye lids. Icy water shuddered from the cracked porcelain sink as I eroded a layer of skin from my face with green paper towels that were apparently impermeable. A glacial stream ran from the towel and down my sleeves, soaking the chewed cuffs of my bottle-green school jumper. The greying pebble of black-fissured soap, refusing to lather, worked its way round and round the mottled plug hole. There was no mirror to check my red-faced reflection, but it didn't matter. I really hadn't been wearing any make-up. Despite this, exhaustion had made my eye lids a shade of heather slate, sat atop the chalk white cheekbones that burned red from cold and exfoliation.

Another night with the curtains drawn wide, from my bunk bed I could see out into the clear and balmy night. In the distance, three red lights glowed, stacked one on top of the other, reaching high into the sky. A radio mast, but it could easily have been a rocket launcher. My two sisters slept soundly, but I was wide awake. I checked the time – one o'clock in the morning. Relief allowed my eyes to close as I counted down to my morning alarm. *If I fall asleep now, I'll get six hours*, I told myself. Screwing my eyes shut, I turned my back to the landing light, its naked bulb left to burn so my siblings weren't scared by the night. I pulled the covers around my head and forced myself to think of sleep. The landing light, ruminating over my shoulder, was somehow louder than any actual sound. It could not be blocked with blankets. Even under layers of duvet-darkness, I knew it was there, effervescently electric. My mind raced with the homework I didn't finish, which exams were next, would I be bullied tomorrow for my trainers, or my acne-marked skin? I shoved a hand under my pillow and grasped a small pink bag. Releasing the drawstring I fished out a small stick doll, brightly coloured wool woven round its lollipop figure. It was one of a few 'worry dolls' given to me, along with a flamingo-pink dreamcatcher – woven like a spider's web with beaded tendrils of feathers trailing beneath it – that was pinned to the wall above Harry Potter's head. Sleep talismans were popular gifts for teens with insomnia. I rolled the doll in my fingertips while staring at the ceiling, my eyes tracing the meringue-tops of plaster patterns, finding features and faces. I checked the time again. Somehow, it was three o'clock and I had yet to sleep a single wink. *Four hours is still OK before school*, I tried to soothe my brain. I returned the doll to the bag, which I replaced under my pillow. Now I felt like I could hear the electricity in the walls of the house, a perpetual humming; or it could have been my own brain. My eyes were glass marbles; heavy weights in my sockets, pinning my head to my pillow, eyelids aflame. Why couldn't I sleep? I was exhausted. It made no sense. My skin prickled and my toes itched, ensnared in my blankets. They were too heavy, suffocating. I threw them off, but the cold air soon saw me reaching to pull them back. I settled for one leg in and one leg out, but still felt claustrophobic. I counted sheep. I counted cows. I pleaded with the Sandman. But still I wasn't granted entry to the dream world. Then, silence was broken. The dreaded, impossibly cheerful heralding of a new day. Though he may have been as black as night, the blackbird betrayed the dusk to usher in the dawn. From somewhere outside my window, I sensed him puff up his chest until it was brimming with song, before letting the notes tumble skyward, as loud as he could possibly muster. He may as well have been sat on my headboard, screaming his beautiful little song into my face. I internalized frustrated tears as that song signalled what I already knew. It was over. It was morning. The face on my wrist read six o'clock.

Sleep has long been regarded a mysterious realm. Even today, we cannot fully explain why we sleep and why we dream. It seems a bizarre evolutionary trait that we absolutely must be unconscious for a good chunk of time every day to survive. We are vulnerable when asleep, unable to be alert to dangers. Sleep has been, and can still be, viewed as a precarious time, when you exist between two realms; not quite with the living, yet not fully on the other side.

Naturally, myths have abounded throughout history when it comes to getting your beauty sleep. The Tudors thought it was too much like being dead to lie down to sleep, so they slept sat upright, to stop the Devil stealing their souls through their mouths.

As a baby that tortured its parents through surviving on very little sleep, had I been born a few hundred years earlier, it could have been a very different story. My nan would often say that I was 'away with the fairies', an explanation for my serene yet sleepless ways. Commonly used today to express a daydreamer, it was once a more sinister accusation. Unlike my parents, earlier humans didn't have the luxury of popping to speak to a GP, their societies were built on suspicion; and a baby who didn't behave or develop in the expected way, could be explained as having been taken by the fairies and replaced by a 'changeling'. Children believed to be changelings could be saved by banishing the fairy through various violent methods, all equally disturbing, often including burning. Academics now think that these changelings were examples of disabled or neurodivergent children, long before society had knowledge these conditions existed. A dark realization indeed.

When visited by the Sandman, we move to another world, a whirling wonderland that melts reality into fantasy, submerging our conscious into the unknown. We never know what we'll experience as we offer ourselves to sleep. Some dreams are vivid and wild; excitement around every corner, before ending in a nonsensical finale; others are so terrifying you jolt yourself awake, only to fear falling back to sleep again, breathless and panicked. Some feel so real you can wake up upset with a person you dream argued with; or make you feel weird about a sex dream in which you saw Piers Morgan dressed like a cowgirl making out with a dungaree-clad Jeremy Clarkson in a hay barn. Or is that a nightmare?

Sleep-resistant children seem to be the cause of many sleep-related folk tales being dreamt into existence, such as the Sandman himself. 'Mr Sandman', the delightful barbershop bop sung by The Chordettes, makes the Sandman sound like a true delight, a gentleman, there to beam a dream man into your lonely life.

The original Sandman, however, is much more akin to the one featured in Metallica's iconic anthem, 'Enter Sandman', with its menacing lyrics. I assume this creepier Sandman was influenced by the band's Danish drummer, Lars Ulrich. You see, in Denmark and other Scandinavian countries, they love a terrifying bit of folklore before bed. Nothing says, 'sweet dreams' quite like the threat of a strange man entering your bedroom, smashing sand into your eyes, and sending you off to never-never land. Grimmer versions tell that the eyes of any child found awake by the Sandman would fall out and be taken to his 'nest' on the moon, where he would feed them to his own children. So relaxing.

In another incarnation, master of horrifying children's tales, Hans Christian Andersen calls him *Ole Lukøje*, (or 'Ole Shut-Eye'). The story depicts (what I think is supposed to be endearing but is actually incredibly creepy) a be-socked man who silently sneaks into children's bedrooms, armed with an umbrella under each arm. He then disarms children by throwing sand into their eyes and cementing them shut, before slinking up behind them to breathe softly on their necks, sending them to sleep. I don't know why Andersen thought this was in any way soothing, nor how he wasn't brought in for questioning, but here we are. When the children fell asleep, he deployed his umbrellas. Good children got the colourful umbrella held over their beds, filling their sleep with joyful dreams. If you were naughty? More neck breathing. Just joking. You got the black umbrella, full of unhappy dreams, if you got any dreams at all. I'd argue the whole premise is, in fact, nightmarish, but this was 1841, when almost anything aimed at children was needlessly hideous. Just by looking at a Victorian doll you run the risk of an eternal haunting. How their children slept at all is a mystery.

Some people dream purely in black-and-white, and men are more likely to forget their dreams in the cold light of day than women. Dreams have common themes: flying, falling, being chased, being pregnant, having your teeth fall out. I became very interested in dreams as a teen, after finding a book on the psychology behind them in the library. Its pages were adorned with whimsical pastel colours detailing theories and lore, moons and stars, hares and saintly scenes; the 'science' was a little sketchy at best, but I was transfixed. My mum would always explain that the 'tooth dreams' we were prone to having were premonitions. I would frequently have realistic dreams where a tooth broke, or I would spit out a molar into my palm. You could allegedly ascertain what the dream was telling you depending on which tooth fell out. The dream book also backed up this theory. For example, if you lose an incisor, it signifies something happening to someone closest to you. The further back into the mouth, the further that person is from you. Mum also assigned different reasonings to whether the tooth was wobbly, rotting or fully fallen out. Theoretically, I can see the parallels. Teeth are an important facial feature, so psychologically the loss of a tooth could be a metaphor for grief or worry. While many may scoff at the idea of prophetic tooth dreams, they are nothing new, having been recorded from a wide spread of cultures across the centuries, with one of the earliest written instances in *Oneirocritica* (or *The Book of Dreams*).

Written by Artemidorus around two thousand years ago, *Oneirocritica* was a 'dream manual'. At a time when magic was an everyday part of medicine, dreams were used by physicians in ancient Greece as a diagnostic tool for a patient's ills. Earlier Greek physicians, such as Galen, believed dreams contained both prophecies from the gods and information on physical ailments. In fact, Galen's success as a physician was even prophesied in a dream, when Asclepius, God of Medicine, appeared to Galen's father in a dream, telling him of his son's future fame. Artemidorus added to Galen's previous knowledge, but dental dreams were his specialty, theorizing a diagnostic tool explaining the symbolic meaning in a tooth's position. A tooth on the left represented a woman, on the right a man; with your molars reserved for premonitions and deep secrets. There I was, in the mid-2010s, being bestowed with tooth wisdom from millennia beforehand. Coincidentally, my wisdom teeth were removed not long afterwards. Maybe the dreams were a premonition, after all.

The ancient Greeks liked a god or two. They had gods for everything, including the night, sleep, and even types of dreams. Nyx was the Goddess of the Night and a child of Chaos, the first of the primordial gods brought into being at the creation of the universe. Before there was anything, there was only night and darkness, represented by Nyx and her brother, Erebus – who was also, awkwardly, her husband. Nyx was feared for her ability to bring sleep or death to mortals; so powerful that Zeus himself feared her and her alone. She rode a chariot across the sky that brought the night in behind it.

The night was revered as a time that could not be tamed, both beautiful and dangerous. From Nyx came Hypnos, Greek God of Sleep, who resided in the underworld along with his mother. On his fabled island of dreams, through which the River Lethe* flows, he sits atop a bed surrounded

* The River Lethe means 'River of Forgetfulness'.

by poppies, lavender and valerian, all plants known for their sleep-inducing qualities. A gentle god, Hypnos wasn't feared. His name, of course the root for the word 'hypnotize', was called upon by humans when in anguish. He bought them relief through sleep, appearing on vases and statues as a young man, arms outstretched, holding in one hand an opium poppy and in the other a jug, pouring the waters of Lethe to bring the relief of sleep to those in need. Sleep was an important time where gods could go cold-calling in dreams, delivering prophecies and future fates, usually distributed by Morpheous, son of Hypnos, and God of Dreams, slipping them into the minds of mortals as they slumbered.

Those blessed with the knowledge of dreams have been held in high esteem throughout the ages, all the way back to the ancient Sumerians of Mesopotamia, whose kings built and rebuilt temples due to the demands of dreams, even to stories in the Bible, such as Joseph. You know, the one with the technicolour dreamcoat.

In *The Book of Genesis*, our Joseph had got into a bit of bother, finding himself in prison. During his time in the cells, his fellow inmates were troubled by confusing dreams. Prisoners were not given access to the priests who would normally interpret them. Joseph offered to read these dreams – any dreams would do – and his interpretations came true. A twist in the tale arose a few years later, when a still imprisoned Joseph was called before the pharaoh. The pharoah is tormented by a repeating dream that no priest had yet solved. Joseph's former inmate just so happened to be the pharoah's cup bearer – proving that it's not just what you know, it's who you know – and Joseph is bought before him to hear the dreams.

> *In my dream I was standing on the bank of the Nile, when out of the river there came up seven cows, fat and sleek, and they grazed among the reeds. After them, seven other cows came up – scrawny and very ugly and lean. I had never seen such ugly cows in all the land of Egypt. The lean, ugly cows ate up the seven fat cows that came up first. But even after they ate them, no one could tell that they had done so; they looked just as ugly as before. Then I woke up. In my dream I saw seven heads of grain, full and good, growing on a single stalk. After them, seven other heads sprouted – withered and thin and scorched by the east wind. The thin heads of grain swallowed up the seven good heads. I told this to the magicians, but none of them could explain it to me.*
>
> – The Book of Genesis, 41

Joseph interprets this dream as prophecies from God, symbolizing a coming drought and famine. He advises the pharoah to prepare and store grain in the seven good years ahead, before the seven bad ones, and this is what was done. The pharoah was so impressed, he made Joseph his second-in-command, showing just how much weight was given to dreams in ancient Egypt. Luckily for Joseph, his predictions came true, and his advice saved Egypt from severe famine.

Dreaming is not just limited to humans. Almost certainly all mammals, birds, and even some reptiles, dream. Scientists believe they have even observed Zebra finches practising their singing while asleep and mimicking the calls of other birds they have heard during the day. You can literally see dreaming in dogs, like my own, who throws some fantastic shapes and has quite the repertoire of sounds, from snores to snarls while sleeping. My dog, Bear, will even 'run' in his sleep, paws scrambling along the floor taking him physically nowhere, but mentally I'm sure he's racing over a beach after his beloved ball.

Some people have harnessed their dreams and turned fantasy into solid fiction. Mary Shelley, famously so. After suffering from writer's block, she had a terrifying 'waking dream' – a 'tale that haunted my midnight pillow' – then woke to find moonlight streaming in through her bedroom window. That dream became *Frankenstein,* widely considered to be the world's first science-fiction novel. Shelley isn't alone, with Robert Louis Stevenson dreaming up the plot for *The Strange Case of Dr Jekyll and Mr Hyde.* Stephanie Meyers' dream of vampiric lovers became *Twilight*. Did the poem *Kubla Khan* come to Samuel Taylor Coleridge in waking consciousness? No, it was delivered in a dream. And Stephen King, one of the most prolific novelists of modern times, exploits dreaming as part of his writing process, calling it 'creative dreaming'.

It's not only plots from pages and film, but other subjects, too. The Beatles song 'Yesterday', came to being after Paul McCartney woke with the tune in his head after hearing it in a dream. Keith Richards of the Rolling Stones, claims that the guitar riff from one of their greatest hits '(I Can't Get No) Satisfaction' was the result of a dream, too.

Scientists turn to a little sleep-boost to work through theories – the periodic table's entire structure was dreamt by Dmitri Mendeleev – but the one that really takes the crown for most impressive dream is Niels Bohr's structure of the atom. That dream won him the Nobel Peace Prize for Physics in 1922. I need some better dreams.

This is a book about 'darkness', so why am I talking about dreams? Because dreams are an essential part of sleep, and to sleep well, we need exposure to proper darkness.

Although debate around dreaming continues to rage on in the world of sleep scientists, it has some vital reasons for taking place; otherwise, why would we dream at all? It may help us learn how to manage our emotions by practising them in different scenarios, and it has been linked to strengthening our memory, helping us recall information when it's needed in the waking hours. Some dreams could be a way the brain tries to organize itself, to keep it free of clutter, and a time for it to analyze events and work its way through problems, hence leading to some creative breakthroughs for those struggling with mental blocks. For some of us who do repetitive jobs or tasks during the day, we can find ourselves doing those things in our dreams. When I worked in a call centre, I would answer calls in my sleep and I know that my husband quite often dreams about whatever book he's read before bed, because he will sleep talk about 'wizards drafting blue'.

Dreams can also be a stress response. During the pandemic of the early 2020s, and the resulting lockdowns, many people were experiencing more vivid and bizarre dreams, and the world headed to Twitter to create #PandemicDreams. Thousands shared their dreams, many of whom pointed to the pandemic as the reason their brain was going awry at bedtime. This isn't the first time global or highly impactful events have caused mass, dream influencing. In 1933, a German-born Jewish woman by the name of Charlotte Beradt, started to be terrorised by nightmares as Hitler came to power. The Nazis infiltrated her sleep as she dreamt of being hunted 'from pillar to post'. She realized she may not be alone in her experiences and set about collecting the dreams of three hundred others in her Berlin community. This was dangerous work, her findings hidden until 1966 when they were retrieved and published under the title *Third Reich Dreams*. The dreams are heartbreaking and horrifying, showing that for those who have suffered trauma, dreams can be a painful reliving of that experience – perhaps as the brain tries to unpick the events that unfolded and make sense of them all. As someone who has suffered with Post-Traumatic Stress Disorder (PTSD), I know all too well that dreams can be anything but sweet. Sleep can be cruel, both by evading you of an evening and by tricking you into reliving your experiences through dreams. Many people with PTSD suffer from recurring nightmares that force them to face the trauma night after night, causing sleep paralysis, night terrors or waking up, drenched in sweat and adrenaline.

In the First World War, many soldiers were diagnosed with 'neurotic depression' and 'stress of service'. These poor men were suffering from what became known as 'shell shock' from the reality of the trenches; what we now understand to be PTSD. One of the hospitals these men were sent to for treatment was Craiglockhart War Hospital for neurasthenic officers in Edinburgh. The hospital is now part of the Edinburgh Napier University campus (where I am currently a student). In the atrium there is a small display of war poetry, written by the patients. William Halse Rivers was a psychiatrist at the hospital who looked at experimental techniques, such as dream interpretation, to address trauma. He wrote a number of works, including, *Conflict and Dream,* which was published in 1923, a year after his death. Rivers believed in patients writing and talking about their dreams in order to work through depression. His patients included two notable poets, Siegfried Sassoon and Wilfred Owen, both of whom refer to being haunted by dreams from the trenches. Rivers' work about the connections between mind, health and dreaming helped change British perspectives on mental illness and how to treat it.

While we may never know exactly why we dream, it certainly matters to our mental health. Our brains must be permitted access to the environments they need, to do what they do best. Dreams are also an important aspect of sleep, which we need to recharge our bodies and minds. In sleep, the brain takes in what it has learned in the day and digests it, creating new pathways and links, to help us learn and retain all this new and (hopefully) interesting information. Sleep reinforces and enhances our problem-solving skills, leading to a happier, more decisive, and more creative brain. If we're not getting our solid eight hours we can open ourselves up to trouble regulating our emotions and moods, risk-taking behaviour and even depression.

Without enough quality sleep, our brains don't function correctly, impairing our abilities and clouding our concentration, and we are less able to fight off infection and disease. Chronic sleep-deprivation has serious consequences, including an increased risk of diabetes, obesity, psychiatric illness, poor overall mental health and even some types of cancer.

Some of us, it seems, are genetically hardwired by our DNA to be night owls rather than early birds.

After glamorously filling a plastic test tube with my saliva and parcelling it in cardboard, I shoved the box unceremoniously into a Pembrokeshire street-corner post box. It quickly found the bottom with a thud. I was sending back my birthday gift of a 23andMe ancestry kit for examination. The posting felt very anticlimactic. I don't know what I was expecting – perhaps that rainbow confetti would fly out of the post-box slot into my face? I now had a long wait for an email to tell me my results were ready. Cue my obsessive checking day after day for eight weeks. I was excited to see what my DNA revealed about me. Were my genes from far-flung lands? What cultural combinations did my ancestry contain? Finally, one March morning, I opened my email to see my results were finally waiting for me. I could barely contain myself, and after forgetting my log-in three times I was finally beholden to the secret history of my genetics. A colourful pie chart denotes the break down, country by country, but mine seemed to be in varying disappointing shades of blue, rather than the kaleidoscopes of colour I had seen for others online. Turns out I am 99.8 per cent North-western European, eighty-nine per cent of which, British and Irish. This was further broken down to essentially, Glasgow and the west coast of Ireland. My ancestors, it seems, did not venture far in the quest to procreate, although the Irish element was certainly news to the Scottish Robertson in me. Still, it was more interesting than my dad's, who is 99.8 per cent British and Irish. At this point I have to assume he's some kind of 4,500-year-old throwback to the Beaker people. The other 0.2 per cent, you ask? Unbelievably, according to his DNA, that matches to Gujarati Patidar, a distinct community from Western India. Dismayed by my lack of cultural inheritance, my eyes scanned the screen to the 'traits' tab. Broken down, line by line, are a list of things you are more likely or less likely to have due to your DNA. Among them, ice-cream flavour preference, eye colour, weight, and to my surprise, sleeping patterns. I had no idea that our genetic make-up plays a factor in determining if we are an annoying and preppy 'morning person' or one of the cool people who stays up all night and sleeps all day.

Turns out, I'm one of the cool people, with a predisposed trait to be a terrible sleeper and a natural wake-up time of 9.03 a.m. Science suggests that if you're a self-professed 'night owl' then you will function considerably less well than your early bird peers, who have no problem matching their schedules to the 'early to bed and early to rise' pattern that our society favours. Experts theorize that humans evolved this way so that someone was always able to keep watch when darkness fell. The night shift for the tribe, security, the bouncer at the fireside. I imagine I've probably got herdsmen in my past that watched flocks overnight. Maybe they lay under starry night skies and wondered what to make of the heavens that shone overhead. Maybe it's these people who passed the night-owl gene to my grandfather – a taxi and truck driver in Glasgow who loved nothing more than a long night drive when the roads were quiet (unless he was up dancing and charming the women with his fancy footwork in the Ballroom) – who in turn passed it to my dad, a man who has always struggled to sleep, preferring an evening walk under the moon to a dawn stroll.

If you're feeling guilty about hitting that snooze button ten times, it might be in your genes. So, blame your parents' parents' parents.

But sometimes, there's other, external aspects stopping us sleeping. There's a concerning upward trend in many of us reporting sleep issues. According to The Sleep Charity, forty per cent of people in the UK suffer with sleep issues.[1] Research organisation RAND discovered that all those sleepless nights aren't just costing us our health, they're also putting us in serious sleep-debt, costing the UK economy

200,000 working days and £41.3 billion in 2016.[2] These issues have become so normalized that most of us don't consider going to get support to sleep.

To be deprived of sleep feels as close to torture as I can imagine, but it has serious consequences. Paying attention is impossible, conversations float around you, reactions are slower, speech slurs and you become clumsy, falling over thin air. When it comes to being tired and doing things like driving, you may as well be drunk. Worldwide, ten to twenty per cent of all car accidents are estimated to be caused by tired drivers, with one in eight drivers in the UK admitting to falling asleep at the wheel. So sleep deprivation can have serious consequences not just for ourselves, but those around us.

Why are so many of us struggling to sleep? There's a myriad of reasons, ranging from stress to smoking, but a global night-time change has crept into our bedrooms and is having a dramatic impact on our chances of a good night's kip. Artificial light at night, or ALAN[†].

How can light impact our sleep?

Light pollution, even in very small amounts, is linked to creating chaos with our circadian rhythm. That rhythm

[†] Sorry, Alans of the world.

is almost universal across all living things. In humans, it's controlled by the hypothalamus in our brains. It may only be the size of an almond, but it is hugely important in regulating functions such as releasing hormones, appetite, emotional responses, body temperature and, of course, our daily cycle. This part of your brain is reliant on light entering through the optic nerves of the eye to 'tell' it what to do. Our eyelids, being pretty thin, means this light penetrates even when asleep.

Our brains release hormones when given cues from natural light sources, and it is the disruption of these hormones that creates such severe effects. Even a loss of one or two hours a night can leave you borrowing from a 'sleep overdraft' that becomes very difficult to pay your way out of.

Throughout the day, and night, your body is responding to these light cues. The light that enters through our eyes is sent to the visual cortex of our brains, which then tells the rest of our body if it's the day or night, dictating what hormones and compounds are released during different stages of the 'cycle'.

Our bodies have evolved over millennia, and our regulatory systems in the brain are a complex result of this evolution. However, it didn't bank on electric lights being spread far and wide in just a mere century. This part of our brain cannot differentiate between the natural light of the sun, or the artificial light of a street light outside your bedroom window.

During the day, when we are awake, a naturally occurring compound called adenosine builds gradually, slowly increasing that sleepy feeling, like sand slipping through an hourglass and collecting in the bottom.

Caffeine is something most of us reach for when an attack of the sleepies takes hold, and for good reason: caffeine essentially blocks your body's adenosine processing. That's why it's not a good idea to ingest anything caffeinated a good few hours before you hit the hay. Personally, I can't drink coffee after 2 p.m. or I'm up into the wee hours, and before I know it I've watched three hours' worth of hairdressing-fail videos, bid for twelve items I don't want on eBay, and struggle to wake up the next day.

While asleep, the body naturally breaks down adenosine to leave us feeling bright and refreshed in the morning. If you've consumed too much caffeine throughout the day, this will 'block' the compound being flushed from the system during sleep and lead to a groggy brain when you wake up, meaning you drink more and more caffeine, and the cycle becomes vicious and you'll never feel fully refreshed ever again.

When it gets darker, our bodies release a hormone called melatonin. This tells our brains it's time to start winding down and prepare to sleep, giving us that drowsy, heavy-headed sensation. We can disrupt this melatonin release by exposing ourselves to light at night; that includes external sources of light pollution, alongside internal sources like TV screens, computer screens, phone screens, basically, all of the screens. I admit to using my phone before bed, but not only is it anti-social for anyone you share your sleep space with, blasting light into your eyes is never going to make you sleepy. It also comes with the added hazard of dropping your phone on your face mid-scroll, which, depending on the handset, could be an effective way to knock you out. So do yourself (and your sleeping companion) a favour and put the phone away, even though that new fringe *is* hilarious. You're only watching something you don't really care about anyway.

Finally, as the sun starts to creep up over the horizon, your brain creates cortisol. This hormone prepares your body to wake up, another very important light cue that's disrupted if you're around bright lights. Imagine what it's like for those of us unfortunate enough to have a bright LED street light shining through the bedroom window? Or a fluorescent sign or billboard keeping a whole street awake? This is light trespass, and something people shouldn't have to sleep with. As we've learned, it can have serious consequences. It also means those of us living in cities, where it is never truly dark, are at much higher risk of chronic tiredness, insomnia and all the other delights that come with a confused circadian‡ rhythm.

So, how do we get a good quality night's sleep to keep our body and mind in tip-top condition? Sweet dreams are made like this; improve your sleep by giving yourself plenty of exposure to natural darkness. If you have a dazzling street light outside the bedroom, ask the council to shield it or dim it. Both are possible. Install the highest-quality blackout blinds you can afford; these will also save you money on heating bills. Resist the temptation to scroll through media at least an hour before bed, and insist on a 'no TV in the bedroom' rule. Finally, invest in a sunset alarm clock. It's the best thing I have ever bought. It doubles as a bedside light that fades through the colours of a sunset to send you off to a natural sleep. Come the morning, it will mimic a sunrise, helping you wake up the natural way instead of being jolted awake by a blaring alarm tone. It really works. I am now fast asleep before the sunset is over and wake up every morning before my alarm, feeling much more refreshed. If it works for me, a chronic, sleep sufferer, it may just work for you, too. Harness the power of light and make it work for your brain and remember, sweet dreams are made of darkness.

‡ Circadian coming from the Latin *Circa Diem*, 'around the day'.

CHAPTER FOUR

A Brief History of Stargazing

The solar eclipse of 1999 was big news, dominating TV for months on end and making the front page of every single national newspaper in the days before and after the once-in-a-lifetime spectacle. It was a frenzied pandemonium across the land. Warnings and emergency planning took place as millions of people were expected to descend on the south of England for prime eclipse viewing, with British Airways even scheduling two special flights aboard Concorde to chase the totality.

People clamoured to find the special glasses that were handed out in schools, papers, newsagents and town centres, as we were warned that directly looking at the eclipse would leave us blind; mine were flimsy, red cardboard with black plastic lenses that I struggled to keep attached to my face.*

* Thousands ignored the advice. Visits to eye clinics and A&E departments were reported throughout the country as people damaged their

Dad walked me miles across beaches and up along the cliff path to the perfectly round mound of earth nestled above the crashing waves of Porth Trecastell. As we clambered over the tufted grasses to the hill, a collection of people came into view – their excitement bubbling over the spent heads of quivering sea thrifts – and down the craggy cliffs, mingling with the sea spray that misted my hair, pasting it to my face. Dad's eyes rolled as the jingling sound of tambourines and other chimes filled the air, swirling with scents of incense. I was perplexed by the scene before me.

We had walked to a special location: Barclodiad Y Gawres, a neolithic burial chamber, its doorway a dark portal on the hillside to another world. At the end of the passage stood two black iron gates, to protect us mortals from slipping into this prehistoric other world. At its mouth, seashells lined the small corridor that lead you into the gloom. Just out of reach were the giant stones, carved with intricate patterns, but it was hard to see them in the womb of darkness. Offerings had been placed at the gates: freshly picked flowers, heads of bright yellow gorse and buttercups bringing the warmth of the sun to the place of the dead. Incense burned, placed on rocks; others had left poems and prayers; some had bound their

---

retinas, unable to resist the urge to take an unprotected peek at the eclipse.

memories and the names of the ones who'd left them behind into tiny scrolls.

I held my arms out, willing them to stretch into the shadows to trace the cold inscribed faces of stone. The incense started to choke so I climbed to the top of the mound. The view was astounding, stretching over golden bays, and out to sea, the faint outline of another land, Iwerddon (Ireland). The sky was completely white. A smooth blanket of unrelenting cloud. A woman stood, twirling, wearing more colours than I knew existed; her arms outstretched, a long white feather clasped in each hand. When she stopped spinning, she handed me a feather, whispering, 'Don't hurt the candle'. I had no idea what she was on about, but took the feather and ran my fingers along its edge, feeling the fibres 'unzip'.

'It's nearly time!' Dad reminded me to put my glasses on, I pulled them out of my pocket. I must have sat on them because they were limper and even wonkier than before. Heavy silence fell around us, the tambourines muted, and everyone stood, despite seeing nothing but cloud, their faces to the sky. I couldn't even tell where the sun should be, but I lay down anyway, glasses balanced precariously on my nose. A weight seemed to settle as the sky darkened, it felt ominous, like the tension before a thunderstorm. Confused birds flocked to the hedgerows and ceased their singing. The entire land was deathly still. Light ebbed away until it felt like we were about to tip into an unnatural night.

No one seemed to move, we collectively held our breath as if we were awaiting a disaster. I felt quite scared, like the world might end. We reached totality without seeing anything, then suddenly the light started to return. A gull announced with a rallying cry that the danger had passed, and the rest of the birds returned their choruses to the sky. The energy around us was strange, lots of people had tears streaming down their faces. We couldn't see the actual eclipse, but to experience the day turning into night was incredible; supernatural. I lay on top of the mound and thought about the people who had built it. I wondered if they would have been scared and what they would have seen without special plastic glasses to help them.

Before electricity lit up our lives, the night skies were a shared spectacle, accessible to all; rich or poor, urban or rural. A wonder to every human on Earth, each person would have been able to step out of their homes and look up into the vastness of the universe. Each star would have been so prominent, even the least inquisitive of minds could not have helped stopping and wondering about just what all those little lights meant, a thousand questions instantly springing to mind. It is no wonder that the stars are so central to many cultures on Earth, old and new. But when do stargazers become astronomers? Is there really a difference? Who was the first to give those fiery filaments names and meanings?

Whether a casual stargazer or those who hold a PhD in Astrophyscial Plasmas or Planetary Physics, astronomy and astronomers have forever been at the forefront of human advancement. There are so many parts of our lives today that we take for granted that would not be possible had someone, somewhere, at some point in time not looked up at the night sky and thought, *Why?*

Astronomy is growing in popularity. During the pandemic, demand for telescopes totally outstripped global supply. When lockdowns came into effect in March 2020, with the majority of the population now unable to travel, it meant that near to no cars were on the roads and planes were nowhere in sight for weeks on end. We had night after night of crystal-clear skies. For many, the lockdowns and the visitor that was Comet Neowise, set them on a path of astronomical discovery. Could it have been other comet visitors, Hale-Bopp in 1997 or Halley's in 1986 that first made you curious about the cosmos?

For some of us, astronomy has been an interest for much longer. Perhaps, like me, the solar eclipse of 1999 is a strong memory for you, or maybe hearing that we had banished Pluto from its planetary status in 2006 made you turn your eyes to the skies.

Or was it Hubble that first captured your imagination? The Hubble Telescope beamed back its first incipient image in 1990, 'first light', before expanding our minds with image after image of the universe ever since. Like Hubble's first images of Jupiter, giving us detail like never before of the planet's Great Red Spot. Photos of comets, planets, moons, distant galaxies, nebulae and supernovae were sent back to Earth in staggering detail that the astronomers of yore couldn't even have dreamed of. The images have since been in the news around the world. I distinctly remember a poster given away free with a newspaper of the Eagle Nebula and its 'Pillars of Creation'. These otherworldly, towering clouds, almost like a hand reaching skywards in velvety hues of blood-red and cosmic blues, surrounded by bursts of golden light, stars being born so very distant from us. The image is so detailed and complex you can lose yourself in the vastness; the how, the why and the where of it all.

Astronomers throughout history have been shining examples of human ingenuity and curiosity, with those who held knowledge of the night sky being important people in our societies, wise men and women, religious leaders and famous scientists. With each generation of astronomers, new knowledge is gained and crafted into the next rung on the ladder of our understanding of our universe.

The first astronomers we hear of today could well be Professor Brian Cox or Sir Patrick Moore. *The Sky at Night* on the BBC was one of my first introductions to 'proper' astronomy. The epitome of a British eccentric, there was not a person in the UK at one time who would not instantly recognize the monocled Moore. He was a much-loved piece of the furniture – an Attenborough for the cosmos – presenting the programme for over fifty years. A self-described 'amateur astronomer', holding no formal qualifications in that realm, he had an incredible ability to take complex happenings and theories such as black holes and eclipses and explain them in accessible, but memorably unique, ways. Sir Patrick witnessed over his lifetime colossal changes in the world of astronomy. When he started his hobby, not a single man-made satellite existed in the sky, and light pollution had yet to plague the UK.

Stargazing Live, also on the BBC, is another highlight, bringing high-level astronomy to the masses, and the universe into our living rooms. It has introduced many of us to places like Jodrell Bank in Cheshire[†] and far-flung observatories on mountain tops, like the Mauna Kea Observatory in Hawaii. Often filmed during challenging viewing conditions in the UK, the show is always entertaining and eye-opening, with presenter Dara O'Brien often having to cover up for cloudy skies with comedy. *But which astronomers came before them?*

† Jodrell's Lovell dish radio telescope was also the setting for a certain Professor's music video for 'Party Up the World'.

Those of us who have even a glimmer of an interest in astronomy will probably be able to reel off some household names. Steven Hawking may be our most famous and world-renowned astronomer, and rightly so. One of the most gifted brains to grace our shores, his books have been a gateway for the masses to get an insight into some seriously advanced, stellar theories. His book, *A Brief History of Time* has sold over ten million copies worldwide. That's enough copies to fill the National Library of Wales nearly twice over, and would need a bookshelf over 191 miles long to hold, longer than the entire River Trent.

In 1931, Georges Lemaître theorized the beginning of the universe starting with a 'Big Bang'; a term which wasn't actually coined until 1949, when astronomer Fred Hoyle used it on a BBC interview to disparage the theory, which he thought was 'irrational'. Lemaître had been using the much less sexy 'Explosion of the Primeval Atom'.

But we need to go further back in time, to find astronomers such as Halley and Newton in the late 1600s, who were both stars in their own time. Isaac Newton was so acclaimed during his lifetime (for the invention of the Newtonian telescope, among other things), that his home, Woolsthorpe Manor in Lincolnshire, has been perfectly preserved. To step into the manor is to be transported back to the days of the Stuart period, a time of significant upheaval that saw the Gunpowder Plot, the Great

Plague and the Fire of London. It was due to the plague that a young Newton had to abandon his studies at the University of Cambridge and head home to the safe haven of the countryside. Here, he continued to build on his scientific works and his chamber at the Manor still stands exactly as it did all those years ago, complete with mechanical sketchings of rays of light on the walls. His scientific thinking is juxtaposed with other wall 'art'; hexafoils, or 'Daisy Wheels' are etched into every window and fireplace to safeguard against evil and witches, reminding us that although science was blooming in seventeenth-century Britain for the great thinking minds of men, it was a very dangerous time to be an intelligent woman.

The infamous witch trials were at their peak during this time, with Lincolnshire having its very own Witches of Belvoir. Three women, Joan, Margaret and Phillipa Flower were arrested, with Joan sadly taking her own life soon afterwards. Margaret and Phillipa admitted to being in cahoots with their 'familiar' (a cat called Rutterkins) and putting a curse on the Earl of Rutland and his family, who fell ill under mysterious circumstances. The two sisters were hanged for witchcraft in 1619. It's hard to believe a society that was becoming so interested in the sciences was one so open to superstition. There aren't many female astronomers that an average person on the street could name, and if having a curious mind could lead to your untimely death, you can see why.

At Woolsthorpe Manor, the fabled 'flower of Kent' apple tree that sparked Newton's Law of Gravity, still grows out in the orchard. It is now in the care of the National Trust and, if you go in autumn, you can take one of the apples from the tree home with you. Little did Newton know that over 300 years later a piece of that very tree would experience anti-gravity after being flown in 2010 to the International Space Station by the astronaut, Piers Sellers, defying the very law he created.

Before Newton discovered gravity, there was Galileo, who gave us the Galilean moons of Jupiter and discovered that our moon had mountains. Before him, Copernicus taught us that the sun was the centre of our universe, not the earth, a huge advancement in human knowledge. Up until then, it was thought that our universe revolved around us (and I'd argue many of us still behave as though it does), and we sat at the centre of everything.

These astronomers take us back to the mid-1400s, 600 years through the ages of what is considered 'modern astronomy'. The truth is, however, that we have studied the stars for much, much longer to try and make sense of the world around us.

Close to 2,000 years ago, in the year AD 185, Chinese astronomers created the oldest known written record of a supernova. Written in *The Book of Later Han*, the observers noted a very bright 'guest star ... the size was half a bamboo mat. It displayed various colours, both pleasing and otherwise.'[1] It was visible for over six months in the night sky. Supernovae are what happen when stars explode, with Asian astronomers also noting a supernova in AD 1054, that was visible to the naked eye during the day for nearly a month! This one is also thought to have been noted by Native Americans by a petroglyph carved into rocks in New Mexico.

Just shy of 3,000 years ago, everyone's favourite ancients looked to the skies and started to write down their musings. It was of course the ancient Greeks and Romans, but it was the Greeks who took the competitive edge when it came to recording theories. Archimedes, Aristotle, Ptolemy and, of course, Plato, were a few among many looking to educate the world about the cosmos.

And 3,200 years ago, another famously intellectual civilisation was potentially using their astronomical knowledge to align their engineering masterpieces, the Pyramids. We absolutely do not know this to be true, but we do know for certain that the ancient Egyptians were very keen indeed on a bit of astronomy. Egyptologists and archaeologists have fought for eons about this, and theories still fly around that the Egyptians used their knowledge to ensure the tips of their pointy pyramids aligned with stars or solstices. The truth is, that while this is a cool theory, we don't have any compelling evidence of this. As humans, we do like to see patterns and meaning in objects that aren't really there, just like the stars themselves and the constellations.

Back further still, to 4,000 years ago, the city of Babylon (now modern-day Iraq) was famed for its culture, beautiful architecture, education and political power. It was once the centre of the Mesopotamian civilisation‡ and home to what some would call the world's oldest known astronomers. It was the Sumerians who laid the foundations for astronomical and astrological knowledge that was later built upon by the Greeks and Egyptians.

We now know that not only were they astronomers, they were much more skilled in the science than we originally gave them credit for. This new knowledge is due to the discovery of five clay tablets, written in cuneiform script – one of the oldest known forms of written language – that have recently been re-analyzed by science historian Mathieu Ossendrijver at the Humboldt University in Berlin. He translated the ancient script to discover they were the intricate workings out of the movements of the planet Jupiter.[2]

Planets were much more fascinating to the Babylonians because they have a habit of jumping around the night sky quite a lot, whereas the stars don't really do much when you look at them from one night to the next. These texts used advanced geometrics in plotting where the planet

‡ You can stop singing 'Rivers of Babylon' now.

would be; a way of astronomical forecasting that was previously believed not to have been adopted until the European medieval period. The geometric principles described on these tablets would be familiar to any physics or maths student studying today. They were using Pythagorean theorem 1,000 years before Pythagoras existed!

But why Jupiter? The legendary Tower of Babel will be known to those familiar with the Old Testament of the Bible. The story in Genesis tells us that the people of Babylon wanted to build a tower to reach the heavens, and in doing so angered God, whose punishment was to confuse the workers by making them all speak different languages.[§] As they could no longer communicate, the tower went unfinished. God then scattered the workers around the world, thus very neatly explaining why we all speak different languages.

It is thought that this myth links to one of the towers of the Babylonian temple of Marduk. Marduk was the Patron God of Babylon, and although they had many deities, Marduk was the most important. He was king of all 600 or so gods and presided over, well it seems, pretty much everything; magic, healing, justice, compassion, fairness, agriculture – oh, and he was the creator of the heavens and

---

[§] In Babylonian the tower is known as 'Bab-ilu', meaning 'Gate of God' or in Hebrew, 'Babel', which is similar to 'balel' the Hebrew for 'confuse'. It is a theory that this is why we say 'to babble', linking back to the story of the workers who 'babbled' and were too confused to speak to each other.

Earth. He became associated with the 'star' of Jupiter, so naturally Jupiter would have been very worthy of noting down in your clay-tablet notebook.

The Babylonians had a vast and varied knowledge that combined what we know today as astrology and astronomy, marrying their celestial knowledge and applying it to divination, prophecy, and their endless quest to please their gods. As with Jupiter, the gods were believed to show their wants, needs and displeasures through the movements in the night sky, communicating to the mere mortals on Earth what they needed to do to appease them. This was known as the 'Heavenly Writing'.

Belief in magic was absolute, prophets looked to the skies to warn of bad omens and trouble ahead. So powerful did the skies seem, it was unquestionable that the movements of planets and stars could not have any physical impact on the natural world. If you have witnessed a night sky with no light pollution, or something such as a comet, eclipse, or meteor shower, I think it's hard sometimes not to feel the same.

Of course, the sun and moon were the most visible, and these too were very important to the Mesopotamians, particularly events like lunar eclipses, which were generally associated with bad omens. It was taken so seriously, that on sighting of one such lunar eclipse, the King was whisked away to a safe place and replaced with a member of the public, who was there to 'fool' the bad omen from the sky.

This poor man got to live like a king for a month before being killed, just to be certain that the omen had cleared. It must have been absolutely terrifying to humans, before we understood the why and how, to see our moon be nibbled away, its light slowly disappearing until nothing was left. Solar eclipses – when the day is replaced with night and all the creatures of Earth fall silent – is spookier still.

So, the Babylonians were the first astronomers. Or were they?

Another discovery from around the same time period is the Nebra Sky Disk. At 3,600 years old, it is believed to be the world's oldest map of the stars. A circular disc made from bronze, it is almost peacock coloured, with a blue-green surface inlaid with gold symbols, one complete circle, what looks like a crescent moon, and a thin semi-circle along the bottom with a scattering of what do look very like stars. They are mainly spread about the disc, but there are seven gathered together, looking suspiciously like the Pleiades.

The disc is impressive, unearthed in 1999 by night hawkers, (metal detectorists who operate illegally, usually at night, at known archaeological sites, stealing artefacts to sell on to private collectors) who inadvertently damaged the object by removing it from the ground. The law eventually caught up with them and reclaimed the disc, and from then on archaeologists have been feuding about its authenticity and the reason for its existence. Found near the summit of Mittelberg Hill, Germany, it has truly captured imaginations around the world since its discovery.

Whatever the true meaning, it is really difficult for archaeologists to find pieces like this and put meaning to objects; left with no instructions and very few clues as to how they were used, because we cannot, as modern-day humans, think the way that humans did in the Bronze Age. It takes years of training to think like an archaeologist and to refrain from putting our own modern perspective into findings. I would love to think that it was used for some significant celestial reason and astroarchaeologists seem to believe it was.

Before light pollution came along, the seven stars of the Pleiades would have been clear to the human eye (now it's usually six or less depending on pollution levels) and academics suggest that the Pleiades would have marked a significant time for our pioneering farmers of the Bronze Age, appearing in the night sky at the very end of the summer, or harvest time.

If you visit the site where the disc was found today, a concave reflective dish has been placed over the find spot, meaning that visitors at the end of summer can look into the earth that held the disc all those years, and see reflected back the night sky its creators would have once gazed upon.[3]

It wasn't just in Germany that Bronze Age findings have alluded to advanced knowledge of the night sky. At the same time in Salisbury, England, the farming communities of old were congregating at Stonehenge. Aligned to coincide with the summer and winter solstices, marking the exact spot the midsummer sun rises and the midwinter sun sets. These dates would have been hugely significant to communities who were at the mercy of the seasons. Winters were harsh and could see whole communities starved if crops failed. The relief of the return of summer must have been elating, to know they had made it through and had a few months of easier living before Earth turned its back to the sun again. The sheer size of the stones, and the effort it would have taken to transport them, speaks volumes about how important these rituals were and continued to be for hundreds of generations. What is even more astounding is that the monument is so significant, it was transplanted there from somewhere else; Salisbury isn't its original home. The stones are bluestones, coming from the Preseli Hills of Cymru over 150 miles away.[4]

There's something about solar and celestial alignments that grip the imagination of our society. Every other week there seems to be new evidence for new meaning to the stones, and at each equinox and solstice the stones are visited by thousands to watch the sunrise and sunset over these ancient megaliths. In 2019, 1.6 million people visited this portal to the ancient world. That's the equivalent of every single person living in Iceland visiting three times in one year, showing that deep in our human psyche we long for those links to live again. They're part of every sinew in our bodies, imprinted in our DNA from millennia ago.

Over time, these stones have slowly told their tales, and they still will as archaeologists continue to unravel the remnants and piece together information, like a giant jigsaw spanning thousands upon thousands of years. They are in some of our earliest stories. In one, Merlin was said to carry them by magic from their original site, believed in folk tales to be Mount Killaraus in Ireland. There they were called the 'Giant's Dance' and link human history all the way back through time to before our existence, when giants were believed to roam Britain.[5]

Ireland, too, has its own equally impressive monuments with alignments to the natural world. Its most notable is the Brú na Bóinne complex in County Meath. Older than Stonehenge and the Great Pyramids, this site is a mammoth of the Stone Age. Its main attraction is Newgrange, a huge passage tomb, an earth cairn sat atop stone walls reaching over forty feet in height and nearly 280 feet across. Sparkling, white quartz dresses the tomb entrance and glitters like the stars themselves. Carved stones stand guard, lining a long passageway to the centre of the mound, a main chamber at the heart with three smaller chambers in the east, north and west, giving the internal blueprint of the site a shape not too dissimilar to a Celtic cross, yet built before the Celts existed.[6] At the entrance to the passageway, a small window exists. Once a year, at the dawn of the old new year, the midwinter solstice sun rises, its beams reaching through this small window, flooding the dark passageway and chamber with sunlight, marking the death of the old year and the birth of the next. Its name is usually translated as 'Palace of the Boyne', but some Gaelic speakers have told me '*Brú*' is also an old Irish word for 'womb'. Bóinn is the Goddess of the River Boyne and was the Goddess of Fertility and Knowledge, so I personally feel that meaning is much more likely. The site is also linked by some to the story of Cygnus, through its annual visitors of a flock of whooper swans, who migrate from Iceland to the site for summer and have done for as long as anyone can remember. This links to an old tale about Aengus and Caer, in which they turn into swans. In the night sky, the constellation Cygnus marks a crucifix in the Milky Way, reflected in the footprint of this ancient site where we once took our dead to return their souls to the night sky.

The Clava Cairns in Scotland are also aligned this way, as is Bryn Celli Du on Ynys Môn (Anglesey). Barclodiad y Gawres, the burial chamber that I was so lucky enough to grow up next to, may not be aligned to a solstice, but its interior has the same crucifix footprint and is now a place that many people enjoy watching a sunset at any time of year, from its perch looking out across the Irish sea. Its carved stones similar to those in Ireland, suggesting the people who built them were connected through time and space, taking with them their customs and cultures with a view as close to 'home' as they could physically be without crossing the sea.

A constellation of these sites lie scattered across these old isles. They're a great place to go and stargaze while thinking about all those people who have come before us and to imagine the world they knew and loved. To see those stars are to know the same stars that your neolithic ancestors did. These places are among the first indicators of the advancement of our civilization, the epitome of 'putting down roots' as we went from being mobile and ruled by nature to becoming settled in one place, making the landscape our home, and starting to transform it to our needs. The night sky was our anchor, it held us in time and place, told us stories and warnings and gave explanations when there was none. If light pollution continues at its current rate, that connection will be lost within the next few decades.

Before Brú na Bóinne was even a twinkle in the eye of the universe, humans were looking to the sky and drawing what they could see. One of the finest examples of this is the Lascaux Cave Paintings in Dordogne, France (now closed to the public due to damage). They were found accidentally in 1940 by a group of teenagers, who while walking through woodland realized that their dog had disappeared. After frantic searching they discovered a hole in the ground and, bravely, the young lads lowered themselves down a 15-metre drop, ever faithful to their hound. Happily, they found not only their dog, but walls spectacularly decorated with scenes of prehistoric animals – hundreds of paintings, depicting animals such as horses, deer and felines, the beasts that the lives of these hunter gatherers revolved around – that these caves had kept secret for thousands of years. They are a time capsule, a period of early human life from over 15,000 years ago suspended in beautiful works of art in shades of ochre, amber and charcoal.

Excitingly, academics studying the art have reason to believe these paintings aren't just portraits of beings of the earth, but also animals from the night sky. These findings have certainly ruffled a few feathers, but further studies by a team at the universities of Edinburgh and Kent have found the beasts not only link to constellations but mark important celestial events that would have rocked their world, literally. One painting shows a horned beast, perhaps an early ox. Next to it, a dying man with a line of direction coming from above the Ox and ending at his feet. The team now believe this depiction shows a comet strike, that travelled through the sky from the direction of the constellation of Taurus. Even more incredibly, the same style has been found in cave paintings around sites separated by thousands of miles and tens of thousands of years, with similar scenes found in Turkey, Germany and Spain.[7]

An even older example found in modern-day Gobekli Tepe, Turkey, is a stone carving on a pillar known as the 'Vulture Stone', which is now thought to represent what would have been a life-shattering comet strike 13,000 years ago. This strike was so significant, it has been blamed as the causation of a mini-ice-age, the 'Younger Dryas period', and symbols seem to depict a shower of comets and their deathly impact. Some have gone a step further and suggested that this place was not 'just another temple', but an early observatory, used to mark and commemorate important events. Carvings are thought to symbolize catastrophe and a huge loss of human life.[8]

By comparing these art works from around the world and using modern technology to date items like the cave art through a method called 'Uranium Series dating'¶, archaeologists have been able to use software to date the paintings and the carvings to significant astronomical happenings, such as the ice-age-inducing comet, and the dates match. This showed that these ancient humans had their eyes on the skies and created ways to track astronomical happenings and the passage of time, making the experts re-evaluate everything they thought they knew about the intelligence of our hunter-gatherer relatives.

¶ Just like radiocarbon dating but using another radioactive isotope.

Now we travel back further still, to the world's oldest known sculpture, the Löwenmensch figurine, also known as the Lion-Man of Hohlenstein-Stadel, which was carved a phenomenal 40,000 years ago. To get some reference for how long ago that is, it's around 5,000 years before the first-known human settlements in Ireland, and around the same time that Neanderthals disappeared. Woolly mammoths still roamed the planet, as did woolly rhinos, giant elk and sabre-toothed tigers. It was a hugely significant period of time to be a Homo sapiens, and our ancestors were battling to survive an ice age. Made from mammoth ivory, it is approximately 30cm, and as the name suggests it's a human figurine with the head of a lion. It is incredible to think this has existed all that time, through dramatic planetary changes. What is even more astounding is that this sculpture is now known to use the same time-keeping system as found in the cave paintings. This is far more complex stuff than just looking at the stars and going, 'Look, pretty!'. This is date keeping, commemorating, the birth of human culture. This is astronomy. Some of the findings of the study published in the *Athens Journal of History*, show that astronomical theory once accredited to a discovery by the ancient Greeks is a whole lot more ancient, indeed.

Yet, we need to look back even further to one of the oldest surviving cultures on the planet – Aboriginal Australians.

Their culture started to develop over 60,000 years ago, and as we will learn later on, it is the longest surviving culture on the planet, holding much knowledge about the night sky, and steeped heavily in the heavens. As Aboriginal culture was traditionally passed on orally, it is nigh on impossible to date the star lore, stories and songs that survive today, but researchers are on a mission to find out more. Aboriginal astronomers could well have been the first on the planet.

When does stargazing become astronomy? It's a hard one to crack. I'm sure a myriad of very scientific definitions exist, but my personal view is that it's the question, 'why?'. I think astronomy starts when we take that step from admiring the stars as a beautiful addition to the night sky and question the reason for them existing, for us existing, to think beyond our here and now, to gaze upwards and inwards.

As I stand on Barclodiad Y Gawres with prehistoric bones beneath my feet and look out upon the ancient light of the Milky Way, it is humbling to know some of that knowledge of the night sky has been passed on through time. As the bright light of the International Space Station emerges from the twinkling depths and arcs across our sphere, I think about all that we have had to learn to get here; how from day one we have always reached for the stars and beyond. Will we still wonder where the next adventure takes us if we cannot see the stars? Let's not find that out. Let's save our stars for tomorrow's world.

CHAPTER FIVE

Women of the Night

Though my soul may set in darkness, it will rise in perfect light;
I have loved the stars too fondly to be fearful of the night.

– 'The Old Astronomer to His Pupil', Sarah Williams

I am running. Running as though my life depends on it. *Thud. Thud. Thud.* Panic raises its hands to my chest. I feel threat at my back. *Thud. Thud. Thud.* My ears echoing the chambers of my heart's frantic beating. I swear if I looked, I would see its shape, cookie-cuttered into my chest, outline pushing outwards through the white T-shirt that chafes at my arms and catches on a bramble net. Red-faced, my oddly cut fringe sabotages my sight. The rest of my hair streams out behind me, laced with beads of sticky buds and grass streamers. I pull my T-shirt free of the brambles' thorny grip, leaving a garland of the finest Fruit of the Loom cotton, strung around a blackberry bauble. I don't have time to waste. *Thud. Thud. Thud.* I am flying again. I am running so fast I am surely the fastest thing on the planet. *Thud. Thud. Thud.* My soles ache as the thin rubber of my shoes repeatedly slaps the pavement. *Don't look back, don't look back, don't look back.*

The atmosphere is changing. The warmth of the sky is draining away, blankets of dark blues are draped heavily across my head. *Just keep looking forward.* I can feel them. Turning. The threat prickles up my back, a looming adversary determined to ruin my life. I can't resist any longer. A stolen glance backwards and I tumble to the ground, a victim of gravity and uncoordinated limbs. Gravel knits itself into the skin of my shins and palms, but I've got no time for tears. I catch my breath and pick myself up, my stinging hands clenched into small, balled fists. I punch the air in front of me, two chubby, ineffectual pistons trying their best to propel a terribly ungainly child on their way. *Ping. Ping. Ping.* It's happening. I can feel them turning. A metallic taste trickles up my throat, a coppery coating splashes across the roof of my mouth. I reach the top of the hill and pause. I see my street at the bottom. Relief. I might just make it.

Suddenly I am hit. I double over as a stitch violently stabs my side. Salt sticks to my face as I force the sea air deep into my burning lungs, my hands on my pot-holed knees trying to forcibly push me to stand upright. Gulls mock me overhead as they glide effortlessly towards the greying sea, sharpened wing tips cutting through the breeze like butter. How I wish I could fly. My envious gaze follows their shapes to the shoreline and I wipe my arm across my face. I can taste salt on my skin. The post in front of me shudders to life, jolting me back to my precarious position. Drawing electricity from the ground, up through its steel-encased trunk, a just-about-audible sound, like a pin hitting the floor, announces both the switch on and my fate. 'Be home before the street lights come on!'

My heart sinks and I start to dig into my brain for excuses. I drag my scuffed and sorry soul towards the front door. *Mum is going to kill me.*

The night has long been the domain of men. Women around the world for all of time have been told to remain indoors once darkness falls. To be a lone woman outside, after dark, breaks the binds of what defines a 'proper woman' in civilized society. We should be good girls, at home with the children or securing ourselves from the heady scents of the night-time air. But the night-time air is just so damn delicious; its perfume is adventure-laced, mystery is at its heart, and fresh, intriguing top notes. But at its base, for women, its scent is heavily laden with fear. That tantalising domain isn't meant for us. By night, men are free to go as wild as they please, with very little to worry them. No one questions where they are going or why, nor does it raise suspicion should they wish to amble of an evening. (Of course, it is going to raise suspicion if you are roaming around at night in a balaclava, with a torch, crowbar and bag emblazoned with 'SWAG', no matter your gender.) Access to the night and natural darkness is a privilege extended to men that most women could only dream of (from the safety of their illuminated beds).

'Should you be doing this job? You know, out on your own in the dark?' It's a question I've been asked a hundred times before. My reply is always the same, 'What will happen to me if I go out at night?' Awkward shuffling almost always ensues. People dance around the answer, unable to give nouns to the dangers they perceive to be waiting in the dark. Many haven't questioned before now what it is they're really scared of, parroting what they've heard before. 'It's not safe to go out after.' No one ever wants to tell me why, although deep down we all know what they are alluding to. Our society deems that as a woman, it isn't safe for me to be alone after dark. What we have created is essentially a curfew on half the population. Any woman out at night is perceived by many to be opening herself up to a wealth of dangers, so we are told to protect ourselves and stay inside.

Mothers and daughters, still having to rush home before the street lights come on, whispering well-trodden conversations with themselves: 'Walk fast. Hold your keys between your fingers. Don't wear a ponytail. Pretend you're on your phone. Wait, no, don't get out your phone. Remember the self-defence lessons. Use your heels as a weapon, but also wear flat shoes, so you can run.'

We lock the front door behind us, close the curtains and block out the night, safe and sound (hopefully) in our artificially lit havens. But is it the dark that's the problem? Or is 'darkness' coded language for an issue that half the population (men) will insist does not exist? Let us defend the darkness and blame this imprisonment of half our population on something else: male violence.

This is an issue that disproportionately affects women and people who are living with disabilities. We all deserve to live in a safe society and one without fear, but once the sun slips below that horizon and the shadows stretch across the darkened streets, many of us step into terrifying uncertainty.

In 2021, the Office for National Statistics (ONS) revealed the results of its first ever study into perceptions of personal safety and experiences of harassment when walking. To the women reading this, it will come as no surprise that where one in seven men felt unsafe walking alone after dark on a quiet street near their home, it was one in two women who answered the same way. Half of all women. Four out of five women felt unsafe walking alone after dark in a park or other open space, and disabled people felt less safe alone in all these settings compared to non-disabled people. All these questions are framed around fear of the dark. Again, the code word. A 'fear' of the dark is justified. Women in the UK are conditioned into a culture of fear and blame. If something happens to us, it must be something we did, something we wore, or we were somewhere we shouldn't be, at the wrong time.

When people ask me if I should be out alone after dark, what they are really asking is, 'Aren't you afraid?' Not of the dark, but of being attacked, harassed, murdered. But no one wants to ask that, because that would give a name to the issue, forcing us to accept it as a genuine problem, this towering issue of violence against women in the UK that society sweeps under the rug as 'the dark'. And, of course I am scared. I have to be. I have to take measures that perhaps my male colleagues would not. It has taken me a long time to de-programme the fear of the dark, to realize what the threat really is and to be confident enough to be out alone at night. It has taken years of steeling myself, telling myself that I will be alright out there, in the darkness, on my own.

Countless nights, a meticulously planned outing, after weeks of waiting for a weather window or special alignment, will go out the window at the last minute. My carefully packed backpack has been left beside the front door, along with my plans for adventure and my nerve. The fear can, and has, got the better of me before. I've broken many promises to myself and not got out of the car when I've arrived at a location; or walked through those woods, because my intuition told me not to. I've cried many a frustrated tear from the inside of my van, looking out at a beautiful place but feeling trapped inside by fear. That feeling in the pit of my very being that tells me it isn't safe tonight. Knowing that if I got murdered in this woodland, my body may not be found and those who care about me will be left wondering what happened. Knowing that by going out alone I'd be bringing judgement upon my actions, scrutinized unfairly by the public and studied by the women who desperately don't want to end up in the same situation, seeing what they could do differently if they were me.

This is the sad reality of woman all around the world. I have lost count of the times men have suggested I'm not a serious astronomer or adventurer because if I were, I would be out on every clear night. We are told that we are paranoid. We are told that we are making things up. We are told it is #notallmen. We shake off the fear, we try to live our lives and step into the night; but then Sarah Everard happens.

Sabina Nessa happens. Maria Rawlings happens. It's a mass gaslighting of women. We are told not to go out at night, but we are also told there's nothing to fear. So, which is it?

This questioning also shifts the blame to me, suggesting that I am somehow irresponsible. That by doing my job and being out late, I am asking for trouble. Of course, every individual is responsible for their own safety, but women should not be made responsible for the potential actions of others – yet somehow, we are. Many of us change our behaviours and routines to protect ourselves from violence and we try desperately to ensure we do absolutely nothing that could be publicly seen as our fault.

According to the ONS, one in five women have experienced some form of sexual violence since the age of sixteen. This equates to 3.4 million female victims. Add to this that over 400,000 women are sexually assaulted every year in England and Wales, and you can see that this isn't all in our heads. UN Women UK found that ninety-seven per cent of women aged eighteen to twenty-four have been sexually harassed and a further ninety-six per cent did not report these situations because of the belief that it would not change anything. These figures are hugely telling about what our society teaches women and girls about what they should accept as the social norm. We shouldn't expect any better. So, we attend our self-defence classes, we walk only through well-lit places (if we go out at all), we share survival tips and tricks on staying alive. We allow our women and girls to live with a gnawing fear, that if we dare step out into the night, we are instigators.

This is wrong. We are denying more than fifty per cent of the population access to the night. This doesn't just mean access to seeing the night sky, it means access to hours of leisure, travel and exercise. A close friend had asked her husband to mind their daughter after work so she could go for a run before it got dark. After being held up in traffic, he arrived home to an upset wife, who was now unable to go for a run as darkness had descended. She told me afterwards, it didn't even occur to her at the time but neither of them questioned this, the absurdity that she could no longer go for a run because it was dark. Her husband took the opportunity to go for a run himself, without a second thought. No one had told him from the moment he could walk that there were rules about where and when you could run, walk, or exist. The inequality of that situation only sunk in days later, as my friend struggled to balance childcare and opportunity to exercise in the darker winter months. Just another barrier to add to the pile. Why shouldn't she have every opportunity to do the exercise she wanted to? Why must women be the only ones to shoulder this fear of the dark?

We are educated to stay safe by walking in well-lit areas. The 2021 tragedy in which Sarah Everard was murdered brought this problem front and centre into the public domain. As her story was shared on social media, the immediate questioning began. *What time was it? Was it after dark? She shouldn't have been walking alone.* The truth is Sarah did everything 'right'. She was walking along well-lit residential roads, had been keeping in contact with loved ones and was wearing bright clothes, but that still wasn't enough to save her life nor protect her from the glaring judgement of the population.

When crimes against women are committed, there's an outcry, and often a kneejerk reaction, with councils installing new street lights to appease public concern. Often, the lighting is installed without much thought, but badly planned lighting can be worse for safety. A Melbourne study (part of the Free To Be Project) worked with young women to map areas in their home cities in which they felt fearful. Thirty per cent of Melbourne participants said they did not feel safe in public spaces at night. This mapping utilized lived experiences of women to inform policymakers and city planners on how to improve the lives and safety of residents. On analysis, the data revealed something quite surprising: locations with higher lighting levels were the ones flagged as creating feelings of fear. On average, when compared to sites where the participants felt safer, the lighting levels were twice as strong.

More and more evidence is coming to light, demonstrating that we are doing lighting wrong. Evidence is building that increased lighting shows no impact on levels of crime. Instead, the opposite is becoming apparent – studies are showing that light can, in some cases, create spikes of criminal activity.[1]

What is going on? In simple terms, lights that are too powerful cause glare, meaning we are blinded to what is around us. They cause deeper shadows and inconsistent lightscapes, our eyes never getting used to the light nor the darkness. It takes at least twenty minutes before any sort of night vision kicks in. These deep-shadow sections may be concealing insalubrious characters and behaviours, but we can't see them because we are, literally, blinded by the lights.

As we get older, the ability of the human eye to adjust to either darkness or light steadily decreases, and we take longer and longer to regain our sight after exposure to bright light. Overuse of inappropriate lighting makes matters worse. But it doesn't have to be this way.

Using the right amount of light, at the right time and in the right place, can create safer, more inviting spaces. Unfortunately, many of those in positions of power to install lighting, such as councillors and local authorities, have very little knowledge around good lighting. Add to this the pressures of budgets and time constraints and they are often bound to install the cheapest, most powerful lights they can find, the end result being something akin to a hostile detention facility. The lights are too white, erasing facial features and whitewashing the world into a sterile space. The naked bulbs practically stab at your eyeballs, flinging unrestricted photons into the night sky, punching you directly in the face. It makes for an uncomfortable night-time environment, not one anyone truly enjoys, and that's before we calculate the impacts on the natural world.

It is not the fault of the councils and councillors. I believe they are doing what they think is best with the resources available to them. It is just a tragic coincidence that this often leads to creating more harm than good. We need to learn that to create safer spaces, we must design safer lightscapes. Every city, town, and rural-planner across the land needs to be educated in what good lighting is. Some communities are taking matters into their own hands, with shining examples of good lighting found in the Welsh towns of Presteigne and Llanarmon-yn-Iâl.

Light or no light, violence against us takes place daily, around the UK. A street light isn't going to save us if a man intent on violence is in the same place as us. All the lighting in the world isn't going to help us. As a society, we need to step up and get to the root of this issue. Only then will women be safe to walk the streets, at day or night.

This is a fight that women have taken to the streets.

Leeds. 12 November 1977. A band of 150 women marched the streets of the city. They lofted placards high, emblazoned with hand-painted stars, a hand clutching a crescent moon and the rallying cry, 'Safety is our right – reclaim the night!' In cities up and down the UK, groups of women congregated. Women in the UK at this point were living in constant fear, as thirteen women and girls in the north of England had been attacked and murdered by Peter Sutcliffe, who became known as the Yorkshire Ripper.

In the wake of those murders, the local police force advised women to remain indoors after dark, unless they were chaperoned by a male. Misogyny was rife throughout the case. Due to many of the victims being sex workers, police were seemingly not as bothered by the murders until the killer targeted women who were 'innocent victims' (in other words, not sex workers). When the advice to remain indoors came from the people supposed to be protecting them, the women of the Leeds Revolutionary Feminist Group decided enough was enough, and the Reclaim the Night marches were born.

Despite four decades passing since then, the same problems persist, and still women take to the streets to Reclaim the Night. After Sarah Everard's disappearance, police knocked on 750 doors in the local area and gave the same advice: 'Stay indoors'.

But why should it be women who have to hide away from the night? Baroness Jenny Jones (a former deputy mayor of London and a Green Party councillor) quite rightly pointed out that instead of policing the movement of women, we should be suggesting a curfew on men. This was said in jest, but it's a valid point. Why should women, once again, be punished for the violence of men against them? This is victim blaming. Yes, I know, it's not all men. But it seems to be, largely, men, doesn't it?

Those of us who have worked in a bar, or done factory, hospital or night shifts know this advice is unworkable. We cannot simply ring up our employers and say, 'Sorry, can't come to work this evening, it's too dark.' Can you imagine? When working as a barmaid, I doubt I'd have been employed too long if I had taken that approach. Every weekend I would be part of the team closing the bar, often not leaving until two or three in the morning. I lived about two miles down the road in student accommodation in the neighbouring town of Trefforest. The bar – one of the most popular watering holes in the Valleys – was The Tumble Inn (and they would certainly Tumble out again, too).

Each Friday and Saturday night, the trains filled with hopes, dreams and hormones would snake their way down the line and pull into Pontypridd, the doors sliding open to unleash the pent-up frustrations of young men from neighbouring towns in spray-on chinos and Lyle & Scott polos. The overpowering scent of Joop! aftershave and stale lager lingered in the night air, long after they had alighted the train. The clatter of heels wouldn't be too far behind. Women in their huddles, linking arms that held handbags containing badly hidden bottles of WKD Blue.

I'd spend my night behind the bar watching the evening unravel, serving copious amounts of Jägerbombs and bottles of Corona; alcoholic anointments to ease the socially inept into awkward attempts at pulling. Every evening I would witness the highs of couples successfully snogging their hearts out on the dancefloor to the romantic soundtrack of Nicki Minaj's 'Spaceships'; and I would witness the lows. The over-consumption to handle the knock backs, or someone who's cup just couldn't contain its cocktail of lust and liquor, shaking and bubbling up until it fizzed over into the inevitable violence we saw every night. The fist fights could be terrifying.

The harassment too. It was like open season; men desperate to catch the eye of a barmaid would blockade our paths, forcing us to engage in hostile exchanges. Men twice my size would pick me up and carry me off, much to the hilarity of their pack. 'Give us a smile', they'd sneer. Or they'd grab at my ponytail or ping my bra strap, like hormonal high-school students, when I tried to ignore or push past them. We were young women, put in harm's way. We were chastized by our managers for letting this happen to us and told to stop fooling around. Eventually, we refused to do that element of the job. Many of us saw our shifts get dropped and our hours drastically reduced. Until one night, there was a 'serious incident' that resulted in a ban on female floor workers, and managers took to wearing stab vests.

The men we would spend all night refuting, or refusing to serve more alcohol to, would often be waiting outside after the doors had been closed. As everyone spilled out into the streets to find their way home, many would hover rather than take the hit to the ego that they had been unsuccessful. A last chance saloon of finding someone to take home. This would result in staff, exhausted from a long shift, having to navigate their way around the stragglers, who would often be slumped in the doorways too drunk to raise their heads, but more often than not just sober enough to follow a woman. Many of us would have to brave the walk home, alone.

In my case, the walk was about thirty minutes if I stuck to well-lit roads, but much shorter if I took the darker shortcut. I'd look down the dark roads and calculate the risks. Of course I should play it safe, but my feet would be aching from the miles I'd already walked after leaving the bar, and the alcohol would have soaked through my cheap shoes, dissolving the stitching of the right sole. As the flap of fabric danced about my foot I'd wonder if I still had the energy to run if I needed to. *Should I just take my shoes off?*

I'd walk quickly, chin down and shoulders hunched, taking up as little space as possible. I'd pull my leather jacket high about my jawline. A taxi would roll by, yellow roof-light glowing above the dark empty seats. My muscles would ache for a five-minute taxi ride, but the temptation to flag down a driver was outweighed by the expense and stories of drivers attacking women. I'd be careful not to draw the attention of the driver in case he thought I wanted him to stop. Up ahead, the taxi would slow down and I'd stop breathing. *Please don't stop here.*

The empty street stretched ahead of me, my heart-beat echoing in my ears. I'd cross the road to slip between parked cars to the opposite terrace-lined pavement. The taxi engine would idle, and I'd steal a glance at the car, seeing only the arms of a man illuminated by the street lights as he gripped the wheel. There's a ginnel up ahead, but would it be a ladder or a snake? My keys felt cold, cutting into the skin between my fingers as I gripped them – Wolverine-style – between my knuckles, just as other women had shown me. The street lights cast shapes; my eyes unable to adjust to night vision due to their glare. I roll the dice. *The ginnel, it is.*

I'd follow the red-brick wall until it disappeared into an empty space. A void. It seemed endless but the taxi was still there. Under my shoulder blades my skin prickles, I draw them down involuntarily and step into nothing. I'd be running now. My shoes discarded. The cold tarmac pushing stones into my skin. I'd be breathing too loud and the copper taste coats my throat, my paces reverberating like an unhelpful echolocation. My tears would smear the lights and my leaden legs refuse to move like I want them to.

Time would stand still, my hands shaking as my keys rattled, unable to find my front-door lock; it would bob and weave its way until it relented, allowing me safe passage inside. I could breathe again.

The sobering reality is, that safety at night is a genuine threat. Employers should either be made to ensure their female staff get home safely, or adjust working hours to keep us safe. We don't want this special treatment, but we don't want to be sexually assaulted, raped or murdered, either. We just want to be able to live our lives. We just want to get home safe. We just want to be safe in the night.

Universities are starting to realize that they have a role in keeping their students safe on walks at night. The University of Oxford has the 'Oxford Nightline'. A number you can call to have someone talk to you while you walk home.[2] In 2023, an app called Routebuddies launched in Wales, which is a way to sort of Uber yourself a group of people to walk home with. And some universities, like Staffordshire University, are setting up safe taxi schemes.

Surprising absolutely no one, this fear of what could happen at night has led to some pretty dismal rates of women in astronomy, astrophotography, astrophysics, and basically anything that means being in a vulnerable position, alone after dark. There are several complex reasons as to why there are fewer females in these roles, such as childcare, access to education, imposter syndrome and more.

However, I would like to argue that if we are denying fifty per cent of our population the opportunity to look into the night sky without fearing for their lives, how on earth are we expecting to see these playing fields levelled?

The International Astronomical Union is a global body representing professional astronomers and astrophysicists. Out of their cohort of 12,514 members, a huge 83.4 per cent of members are male, meaning women make up just 16.6 per cent of memberships. A clearly galactic gulf of divide. A paper titled 'Closing the gender gap in the Australian astronomy workforce' by Lisa J. Kewley from the Research School for Astronomy & Astrophysics, Australian National University, found that if nothing changes in how we encourage and, importantly, support women in astronomy, it would take until the year 2080 before professional female astronomers would make up even a third of all astronomers in Australia. I've never had the good fortune to visit Australia, but I know just how pristine their Southern Hemisphere skies can be. Seemingly access to darkness is a limiting role for women in Australia too. Lisa's paper states:

> *In astronomy, the fraction of women at the senior levels in academia remains critically low worldwide; the fraction of senior women in astronomy in the US, Germany, Canada, Australia, China and the UK has remained at or less than 20 per cent for several decades.*

There are a number of reasons for this, including gender bias, women being made to feel uncomfortable by male peers – even things like access to time on telescopes, with men being given priority over women in the same field. Not only do we have a wealth of social bias to overcome to get out into the dark in the first place, the situation is then compounded for those women pursuing astronomy as a career.

Even astronauts don't get a break. When America began its search for the first astronauts, NASA naturally fixated on their male cohort. One man, Dr Lovelace, however, and a small number of his colleagues, had a sneaking suspicion that women might make better astronauts than men. On average, they weigh less, are shorter, need less food and consume less oxygen. When you are wanting to catapult humans into the great beyond, weight and fuel consumption become critical. It made perfect sense on those grounds, but were women physically capable of the gruelling demands of space travel?

Had they ever met a woman? Of course they were capable. A now legendary group of women who formed the Mercury 13 in the 1960s, were put through all manner of physical, medical and psychological testing. This trailblazing crew juggled all the responsibilities of women at that time* to ace pretty much everything that was thrown at

* One, 39-year-old Janey Hart, was a mother of eight!

them, surpassing the male astronauts in almost all tests. It turned out that women *were* better suited to being astronauts. The only thing holding them back? That fragile old thing called the male ego. It was deemed simply unthinkable and damaging to the American image if a woman was sent to space before a man. Jerrie Cobb, Lovelace's first recruit, was an accomplished pilot before she had even reached adulthood, thanks to her pilot father, Lt. Col. William Cobb, who encouraged his aspirational aviator daughter from an early age.

When the women of Mercury 13 were stopped from going to space, they lobbied Vice President Lyndon Johnson. Cobb was quoted as saying,

> *There were women on the Mayflower and on the first wagon trains west, working alongside the men to forge new trails to new vistas. We ask that opportunity in the pioneering of space.*

They never made it. They were told their feet were to remain planted firmly on the earth and in the home.

Interestingly, the Russians didn't hold the same views, launching their first female astronaut, Valentina Tereshkova, into space in 1963. It would take the Americans another two decades until they did the same.

This wasn't the first incidence of women defying men and taking to the skies. It is 28 June 1942. Hitler is closing in on Moscow and the Red Army of the Soviet Union is in trouble. Under cover of darkness, a fleet of flimsy, open cockpit Polikarpov Po-2 plywood biplanes are making their way to the Nazi headquarters. The air is so cold, the pilots are fighting off frostbite. Their badly fitting, hand-me-down uniforms give barely any protection from the biting cold. The planes are so small, they have no radios or radar, not even parachutes, in order to keep the weight down. Suddenly, a red flare illuminates the sky and the wings of the aircraft in front. The target has been located. A spotlight streaks into the night sky. It is time. As Nazi planes fly up to meet them in air-to-air combat, these daring pilots split. The leading planes lure the enemy aircraft into a wild chase across the night sky, while the rest of the pack idle their engines, turning themselves into near-silent gliders. With great skill and stealth, they dive in total darkness towards their target. The only noise to be heard is the swoosh of air over their canvas wings, like a witch's broom in the night. '*NachtHexen*!' the cry goes up –'The Night Witches are here!'

The 588th Night Bomber Regiment was the first all-female unit allowed to fly combat missions, not just in the Soviet Union, but in any of the nations in the Second World War. Born from desperation rather than any consideration of gender equality, Stalin approved the request of Marina Rashkova, already famous for becoming the Soviet's first female navigator, to recruit an all-female squadron. Inspired by the letters from hundreds of bereaved women who begged to play a part in their country's defence, Rashkova set to work. She received over 2,000 applications from Soviet women, of which 400, between the ages of seventeen and twenty-six, were selected to take to the skies. What took their male counterparts two years to learn, the women did in six months. Not only were the ancient planes made of highly flammable materials, they only had capacity for a single bomb under each wing, and no guns. The weight limits meant they had to fly low and slow, leaving night runs as the only viable option. They faced sexual harassment and misogyny, with their fellow personnel finding it shameful that these wily witches were doing such 'unwomanly' work.

Despite everything they faced, these female pilots were hugely successful, carrying out multiple bombing runs each night and going on to be one of the most highly decorated units in the Soviet forces. They harassed the German military, their swooping biplanes leaving the Nazis terrified and in disbelief that these daring tactics were being used by women. Rumours abounded that they were a fleet of criminals, given drugs to help them see in the dark, or, of course, they were using witchcraft. In reality, they were just badass, fearless women.[3]

Scientifically minded women have been, throughout the ages, marginalized, berated, humiliated and deemed insignificant. Any woman, for myriad reasons, could run the risk of being labelled a heretic. The hunt for heretics were what eventually led to the hunt for witches. In Scotland, a great many women lost their lives to the noose and to the flames for their perceived association with darkness. Women who helped with midwifery (midwife literally meaning 'with-woman' at childbirth) were often accused of witchcraft.

Margaret Bane lived in Aberdeen and was well respected for her skills in easing the pains of childbirth. Many women called on her to be at their side when labour took hold, including nobility. She was accused several times of witchcraft, but in 1597 her midwifery skills were brought against her as evidence of her meetings with the Devil and sorcery. She stood accused of murdering a man by transferring the pains of labour onto him, compounded by witnesses stating she paddled in a loch (how dare she) and did not exchange greetings with a man, who later died. Poor Margaret, clearly highly skilled in caring for women through the most vulnerable and dangerous time in their lives, was found guilty. She paid with her life and was burned at the stake on 24 April that year, on the hills of Aberdeen.

Women who run with the night run the risk of bringing their reputation into disrepute. 'Lady of the night', was once upon a time a name for the moon; but today you may be familiar with the phrase as an old-fashioned euphemism for 'prostitute' – itself a derogatory term for sex workers. However, in Victorian England, even the word prostitute extended to cover far more than a woman who sold sexual favours, including a woman who lived with a man she was not married to; a woman with children born out of wedlock; a woman who looked or behaved in a way men deemed unacceptable, such as daring to walk the streets of an evening, no matter their intentions. Being a 'difficult woman'[†] was sometimes all the evidence Victorian doctors needed to lock troublesome women away in their archaic-named lunatic asylums.

Even the word 'lunatic' has its etymology chained to the shackles of the dark and dangerous night. Luna, Roman Goddess of the Moon, gifts her name to the etymological foundations of the Latin *Lunaticus,* literally meaning 'moon-struck', stemming from the myth that moon-gazing could leave you mentally impaired. (Or a werewolf.)

Full moons have long been the answer to humans behaving badly; to this day people hold beliefs that a full moon has the power to influence us water-filled beings into behaving erratically, bearing the blame of increased rates of road accidents, hospital admissions, and violent and criminal behaviour. Police in the UK have even been known to add extra staff to shifts during a full moon, to

[†] read – refusing to do husband's bidding.

cope with us poor, moon-struck beings. We may be eighty per cent liquid, but the moon bears no gravitational bearing on our internal fluidity. For the same reasons, as lakes and rivers not having tides, the amount of water is just too small for the moon to have impact. So, if you're being a dick, you don't blame the moon. (And while we're on the subject, your dickish behaviour can't be blamed on being a Sagittarius, either.)

By the late 1800s, women were feeling more in tune with themselves. They were rebelling. They were organizing. They were *thinking*. And the men did not like it one bit. So, with all their newfound power, they took to the shears and cut their fringes. A 'lunatic fringe', to be exact. Men hated these haircuts so much, they told women they looked insane; and a new craze was born.

One of those women exploring their new freedoms (but not fringes) was Thereza Dillwyn Llywelyn. Her father, Lord Penllergaer, owner of a sizeable country estate on what was then the rural outskirts of Swansea, was a well-renowned photographer, though not as talented as his younger sister and Thereza's aunt, Mary Dillwyn, who was one of the world's first female photographers and certainly the first notable female Welsh photographer. There's no doubt that Mary inspired her niece. But not only was Thereza interested in photography, she held another passion: astronomy. Thankfully, despite Victorian sensibilities, her father seemed to reinforce this passion and for her sixteenth birthday he built her an observatory in their mansion's grounds. Talk about a sweet sixteenth! Combining her two passions, Thereza worked together with her father to develop a technique for taking one of the earliest photographs of the moon – and the first photo of the moon from the UK.

Women deserve access to the night. They have a right to roam beneath the light of a nearly full moon, or simply look up at the night sky without feeling fear. We cannot deny half of our population access to the night any longer. It is time society dealt with the societal issues that lead to male violence. It is time to demand safer, Dark Sky friendly, street lighting, to light the path to an accessible night for all.

CHAPTER SIX

Silent Night

We sat and listened as the sun sank into the deep. The air, cooling rapidly, washed clean the redness on our faces. The cries of waders carried across the silvering tidal flats; curlews bubbling, hidden from view as small flocks of knot hugged the curve of the shore. Gulls called out to one another as they flew into the fading rays, wings skimming the small breaking surf on the last waves of the day. Standing slowly as the warmth trickled away, the evening chill soon made itself known in our bones. We walked home through twilight hues. Starling shapes pulsated, their murmuration giving a heartbeat to the sky. So many feathered bodies welcoming the night. 'We'll walk the long way home,' Dad said, who could never tear himself away from the birds. It was as if he was attached like Velcro to the crisp evening air, having to slowly peel himself away. Along the north side of the lake, the reeds felt ten feet tall; reeds turned into tower blocks, alive with the whirring and chirring of warblers, tits and finches, all settling in for the night and exchanging stories with their neighbours. The reeds let out creaks and groans under the weight of their tenants, like an overloaded ark. We reached the fields, where the donkeys were grazing, tails flickering, annoyed by the insects that were descending. To their rescue came first the swifts, screaming, shearing the air with their razor wings. 'There's Sirius,' said Dad, waving an arm to the south of the sky. When I asked how he knew, he shrugged. There in the early evening shone a bright light, suspended just over Yr Wyddfa's (Snowdon's) mighty peak. 'Now, here come the bats,' he added. To my delight, he was (as usual) right; six or seven shadowy shapes now darted acrobatically along the hedgerows. I heard their high-pitched calls in my young ears and paused to watch these incredible hunters in the quickly fading light.

It is a common misconception that the night is quiet. It's true that it's becoming quieter as so much of our wildlife disappears; but if you take a walk in the evening or even the dead of night, you'll hear that there's a lot that goes bump, flutter, snuffle and screech in the night.

More than sixty per cent of the UK's wildlife depends on darkness to survive.[1] This figure excludes our insects and marine life. For so long, and in so many ways, we have been destroying their habitats, whether knowingly or not. When it comes to light, our inconsiderate usage has been silently impacting on our wildlife without many of us realizing. At one time I considered night-time to be when wildlife perhaps got a break from human meddling, unfortunately it couldn't be further from the truth . . .

Just think about all the obvious ways we have been pushing back their habitats; sweeping swathes away from our neat and organized lives with no room for the clutter and chaos of nature; flattening trees and hedges to make way for tidy rows of identical developments with lifeless drives and plastic lawns. Agricultural monocultures, ploughed, sprayed and grazed, borderless, with hedgeless edges. Rivers and streams tamed, straightened and polluted. Our oceans are smothered in plastic, ghost-like bags that float and catch themselves in its airways.

There are so many battles that nature is fighting, it is not surprising that this one has gone unnoticed for so long. Light pollution. The biggest environmental disaster you've never heard of.

We have been silently pouring into our dark wild places and spaces with unrestricted light. Light can reach its rays far and wide from its intended home and usage. Humans have shone artificial light into every corner, putting an eternal spotlight on empty homes that once housed some of our most beloved species.

Externally lighting houses and other property, like our cars, gardens and garages has created an ironically unseen problem. When we are asleep and tucked up in bed, these lights that burn bright all night have hidden in their glare an ever-growing disaster for our wildlife, destroying habitat that was once safe in the shadows.

Perhaps the most famous victim of light pollution is the bat. The UK's only true flying mammals are the bane of every property developer who has ever heard the words 'bat survey'. To many they are a nuisance, seen as holding up building projects and ruining development plans. Some people are even fearful of them, thanks to their unfair depictions as blood-sucking virus carriers. But our bats are a valuable 'indicator species', meaning they are vitally important when it comes to understanding the health of our overall environment. Since 2008, bats have been one of the species used by the government to help measure their goal to halt biodiversity loss. If our bat numbers start decreasing, it means we have big issues somewhere else in the ecosystem, which, of course, would affect our species, too.

Bats live in a wide variety of habitats, from farms, woodlands and abandoned buildings to even our most urbanized areas and homes. Outside of the UK, bats are a crucial pollinator, just like bees. Seen as pests to some, they are an effective pest-control unit. A single pipistrelle bat needs to eat around 3,000 insects in a single night. So next time you're being eaten alive by midges in the garden, why not employ your neighbourhood bats?

One of the biggest ways we negatively impact our bats is through improper use of light. While not all of the UK's eighteen bat species is *negatively* impacted by light, every species is impacted in some way. They can't always adapt to overcome the change light brings. Bats are highly evolved to hunt at night for several reasons, one being that they are less likely to be targeted by birds of prey. This, however, means they have developed ways to live in a Dark Sky that is totally disrupted by light pollution. They have poor eyesight (though not blind, as widely believed) but more than compensate with their echolocation. At the start of each evening, bats will 'light sample': partially emerging from the roost to see if the conditions are right before taking the great leap out into the night to begin their foraging. Even having a light on partially into the evening will stop bats leaving their roosts, meaning they could miss vital foraging time, when the insects they prey on are most active.

As many species will avoid light to protect themselves from being preyed upon, sometimes light can block flight pathways, closing off travel corridors to hunting grounds vital for their survival. Artificial lighting also attracts insects. We've all left the window open and a light on over a summer evening, and woken to a bug bonanza on the ceiling the following morning. What happens when lights attract insects is they are removed from these hunting grounds and become inaccessible to bats. A couple of species take advantage of this using the light as a prey magnet, but for the majority it means a huge chunk of the ecosystem has been shifted out of reach, leaving bats to go hungry. This has led to a decline in some of our most vulnerable species such as the greater and lesser horseshoe bats. The ones who take a chance to hunt in the light, risk becoming prey themselves, with recorded examples in our cities of peregrine falcons successfully attacking bats under street lights.

I have seen some really thoughtless light installations where lights were being shone directly onto the opening of bat roosts, effectively entombing any bats that were inside and destroying a roost for those outside, removing even more of their valuable and ever-shrinking habitat.

Of course, if you're one of these strange people who doesn't like bats – don't go getting any bright ideas about using light against them. It is a criminal offence to disturb, remove or cause harm to bats and their roosts, and this is inclusive of using light as a deterrent.

Sat on the hard, wooden floor of my primary-school hall, I was excited to see a local bat group were there to give an assembly. I itched with anticipation as I noticed small timber boxes alongside the information stands. When they asked for a volunteer, I threw up my hand so quickly that I nearly knocked out the child next to me. I fell over myself to get onto the stage. An older woman, dressed in a green raincoat and a shocking-pink scarf, asked me to hold out my hand. Carefully, with her back to me, she opened the lid of the box. She turned back, hand clasped shut. I rolled up the chewed and stretched-out sleeves of my school jumper and held my hands out towards her. I nearly burst with joy as I was handed the most beautifully delicate common pipistrelle. It had been rescued as an orphan and raised by hand, sadly that meant it was unable to live in the wild. I was besotted by the intricate little creature in my cautious grip. Its perfectly crafted, leathery wings folded in against the dark golden fur of its body, two small claws hooked over my fingers as I tried not to clutch it too tightly.* I was struck by how fragile it felt as I studied its comical face and oversized ears. It felt lighter than air. Like I could unfurl my fingers and it would just hang there, suspended. Almost certainly the best day at school ever.

I grew up near a lake, its footpaths worn down frequently on walks with Dad. On warm summer evenings, just as the sky lost those warm tones of gold, I remember the shapes in the sky turn from the swooping swallows and swifts to the fluttering of bats. Zigzagging along the hedgerows of blackthorn and dog rose that framed the fields,

* Pipistrelles weigh less than a 2p coin.

these trees became long banqueting halls, teeming with a feast of mini beasts. The campsite nearby had stables and old barns, whitewashed and a little tumbledown. The loose tiles and weathered, lichen-splashed stone gave bats and other animals the perfect places to sleep the days away.

It's in one of those buildings that I currently live. Now developed into a small home, I admit when I realized that this dwelling must have been one of those roosts I was wracked with guilt. However, I am pleased to say it is still home to a number of *Pippistrellus pygmaeus* (soprano pipistrelle) who live in my roof, thanks to the landlord's sympathetic building approach, which has left nooks and crannies available to the wildlife. I'm fortunate enough to work somewhere with lots of wildlife gadgetry, including an echometer bat recorder. When I first used the echometer I was expecting some hugely complex bit of kit. I was perplexed when a box the size of a matchbox was given to me. Impressively small, it simply plugs into the bottom of your phone. You load up the app and like magic, your boring smart phone is now a bat recorder! The recorder not only picks up the bats' ultrasonic echolocation call as a visual sound wave, it identifies the bat species, even showing you on a map the flight path of the bats. It's proper entertainment to wait for the light to drop enough and see them exit the roof, one by one. Walking along the hedges as they fly around above you, the echometer instantly lets you know exactly what species is accompanying you on your evening walk.

You don't have to live in the back of beyond to encounter wildlife at night. In fact, I've seen more wildlife at night in towns and cities than I have in the countryside. A lot of our species have developed excellent night vision, giving them the upper hand over their prey. Darkness is essential for them to hunt, to sleep and to live. Human and animal nightlife can collide in wonderful chance encounters.

One of my favourite encounters tells a tale of my misspent youth, dragging a friend from one of Holyhead's finest eateries. Cheesy chips in hand, the sound of thumping bass and laughter fading into the background, we navigated our way home from the local nightclub[†]. My friend, Sarah, was complaining her heels were hurting as we clung to each other to keep both upright and warm – as no self-respecting teen wore a coat on a night out, even if it was March and you were wearing hot pants. We stopped so that she could remove the offending stilettos when we heard a growling noise in the shadows, between two terrace blocks. There was a rustling and a clanging of glass and tin as someone's old Dolmio sauce jar rolled into the road, noisily dropping from the kerb. We panicked. Expecting a dog (or worse) to come flying from between the bin bags, we were ready to

[†] affectionately known as 'The Pit'.

abandon our precious cheese-topped cargo and run barefoot through the streets. As we moved swiftly (probably not incredibly so) we were nearly bowled over by something much more exotic – a badger! I'd never seen one before and was left frozen and silent in its presence.

It was much bigger than I thought badgers were, peering along its striped snout, it had a regal air about it, king of the town's back streets. Suddenly, it stormed off into the road, intermittently illuminated by street lights casting stripes along its stripes. It stopped briefly to look back at us (its body language definitely judgemental) before trotting off in the direction of some allotments. Sadly, I am yet to encounter badgers in the countryside, but I have seen several in urban settings. They're excellent animals and I am jealous of the people who are lucky enough to have gardens to which badgers, and other nocturnal creatures like foxes and hedgehogs visit.

The sounds of the night are unforgettable and often chill you to your core. An unknown creature calling out can set the mind and pulse racing. I'll never forget the first time I heard a fox cry; I was sure it was someone being murdered, blood curdling at the harrowing sound that broke the night in two. It's often the foxes that really scare people when I'm giving talks or tours. It can take a while to convince people that their lives are not in any danger as I try to remove their fingernails from my skin. Foxes, too, are something I have seen more frequently in urban settings, rummaging through bins, eyes caught in headlights or seeing a bushy tail out of the corner of my eye just as it slinks away.

Living in Neath, I rented a bungalow, which was rather hypocritically sat at the top of three flights of steps that were cut steeply into rock, taking you past three awkwardly shaped little parcels of garden. You needed serious hill-fitness to traverse up and down. One night I struggled up these steps, carrying a back-breaking amount of shopping slung under both arms, the plastic handles cutting into my shoulders, tins bashing against my knees. Cursing my lack of fitness, I fumbled for my keys. Suddenly, I felt eyes on me. A curious fox sat in the midway garden looking back at me, quite serene. The most ethereal face, looking directly into mine. Not a flicker of concern crossed its glowing amber eyes, lit by the street light behind me. I clumsily went for my phone in the hope of getting a photo, looking down for the briefest moment only to realize the fox had silently, effortlessly, vanished like smoke. I still question whether that actually happened, or the lack of oxygen had made me hallucinate.

Foxes seem to be adapting to an artificial night, the bright, city lights making it easier for them to hunt prey or, more likely, find food to forage. They are becoming increasingly visible in urban areas, with some academics suggesting they're domesticating themselves in a similar way to cats and dogs. The University of Glasgow led a study that showed urban foxes have developed smaller brains and a different snout-shape better for foraging.[2]

They're also becoming much less fearful of humans, with some even being bold enough to snatch sandwiches from family picnics!

⁂

On those first, milder, mizzly spring nights, where the rain settled across my face like a veil of vapour, the dark country lanes around my home became besieged by a battalion of frogs and toads. Every year they marched back to the spot they were spawned, looking to mate. As headlights swept the roads, I realized that the lumps and bumps were amphibians. Trapped on the road by kerbs and attracted by lighting, they didn't stand a chance in those trenches, funnelled into a flattened fate as no one stopped to save them.

Around the UK, volunteers don hi-vis jackets and take to the roads on toad patrol. Like Day-Glo shepherds, they usher, scoot and scoop up these beautiful beasts, clueless as to how close they came to a terrible fate. Some come quietly, sitting squatly in the palm of your hand, shiny eyes and skin glistening under your headtorch lamp. Maybe this isn't their first toad-patrol experience and they're grateful of the giant gloved hand that plucks them from obscurity and into a bucket with their comrades. Others give a struggle and squirm; you cup them carefully, willing them to cooperate. They're collected in buckets and released on the other side of the road so they can make their way to the pools, ponds and lakes of their birth unharmed. The frogs leap away from you, glorious, long-limbed legs pushing into the damp mulchy ground, propelling them with surprising grace onwards to their romantic plans.

The toads are less rushed. Built like squat tanks, they trundle along with much less haste. Their rough and warty skin more akin to bark than flesh, blending expertly into the undergrowth the minute you take your eye off them. You send them on their way, hopeful that soon you'll find the evidence of their success; the frothy, fish-eyed masses of frogspawn bobbing among the reeds, or the translucent jelly pearls of toad-spawn lines, threaded carefully among the emerald-green pondweed. Each black iris, a prospective new life, is left to fend for itself, until it's ready to join the ranks and cross the battlefield roads the following year. Roads are built without much consideration to the animal migrations that need to take place year after year across their tarmac flats. As a result of this, a staggering twenty tons of toads are killed every year on British roads.[3] Toads appear to be drawn to lights, another cause for their demise on the roadsides with street lighting proving too much of an alluring presence. A study in Ohio has proven light pollution to be so damaging to our amphibian friends that toads subjected to constant light had their growth stunted by at least fifteen per cent, and it stopped frogs laying their eggs altogether.

We really do have some curious creatures that thrive at night. Another favourite of mine is the elusive nightjar. It uses the lunar cycle to guide its migration from sub-Saharan Africa to the UK. By day, its mottled plumage resembles bark, camouflaging perfectly into the forest floor. These birds love woodland, especially commercial plantations. It's unlikely you'll spot them during sunlight hours, but as dusk falls you might get lucky and hear the unmistakable 'churring' call of the males. This unusual call was what first made humans suspicious of this crepuscular creature. Known colloquially as the 'corpse bird' it was a harbinger of doom and a bad omen. People believed the distinctive calls carrying through the darkening sky were in fact the cackle of witches, hiding in the bushes, preparing to steal their livestock and children. Their scientific name, *Caprimulgus europaeus*, belies another superstitious attribute. Roughly translating to 'goat sucker'; in the days when we believed that birds migrated to the seabed (or the moon), we also believed nightjars were stealing milk from our goats. This was mainly due to their vicinity to livestock, but they were after the insects that livestock attract. This accusation was also cast at our hedgehogs, with Henry

VIII being particularly harsh on our garden friends, paying people to slaughter as many as they could; it's a wonder we've any wildlife left at all.

Perhaps the biggest losers when it comes to light are our insects. Often (and ashamedly), we overlook our insects; denying them the respect and admiration they deserve, for they do a great deal for life on Earth. In fact, they are the foundation for all of life on this planet. They pollinate our plants which in turn gives us the food we need to survive. They care for our soil, fertilizing it, giving it the nutrients that sustain the life that grows from it. They also feed many life forms on this planet, and without them, well, we are all dead. They're worth a whopping £500 million to the UK economy just through their powers of pollination alone. To put this into perspective, that's more than Britain's Royal family contribute to our economy. It would also enable you to buy six flights to the International Space Station with the Russian Space Team; about twenty-seven of the world's most expensive car (the Bugatti La Voiture Noire), or two-and-a-half Hubble telescopes[‡].

Insects are in huge decline, with talk of the insect apocalypse being the next great extinction event. Spoiler! It won't end well for us, either. Globally, our insect populations have been crashing, with a terrifying forty-one per cent global

[‡] Not inclusive of launch costs.

decline in the last decade alone.[4] This is down to a number of reasons; changes in farming practices, use of pesticides, and a loss of habitat. However, it's now believed that one of the key driving forces behind our insect apocalypse, is in fact, light pollution.

We all know the saying, 'like a moth to a flame', but sadly it's true of any light. Moths are drawn to light bulbs in our homes, you hear them flapping frantically against the hot bulb until they die. Our street lights are no different. Huge swathes of all manner of winged insects will be found flocking to external light sources, potentially confusing it for the natural light of the moon. Most will exhaust themselves, eventually collapsing and dying. Others find themselves an easy target as their predators take advantage of a street-lit dinner. This destroys ecosystems, removing food from the very things that they need to survive. They don't pollinate the crops and flowers that need them. They fail to find mates; therefore, they don't reproduce, having an immediate and decimating impact on our insect populations.

Moths are just boring, dusty butterflies, aren't they? Wrong! They're incredibly varied, some more spectacularly coloured and camouflaged than their garish day-flying cousins. There are a mere fifty-eight butterfly species in the UK, wholly unimpressive compared to the 2,500 moth species. Over one hundred of these are day flying themselves, such as the cinnabar moth (with the orange-and-black caterpillar commonly found on ragwort). Some species have evolved so that they can hear a bat's echolocation and take evasive flying manoeuvres. They have some spectacular names, just a sample few; the Barberry Carpet, Brindled Beauty, Elephant Hawk, Maiden's Blush and my favourite, the Death's Head Hawk Moth. A total gothic fantasy, this moth is famed for, as the name suggests, a pattern resembling a skull on its head. Naturally, thought to be a bad omen in more superstitious times, it has an amazing skill; it can imitate the squeaks of a queen bee[§], duping the workers and allowing itself entry into even the most exclusive of beehives where it will feast on the honey.

It is our unsung hero, the humble moth, who is suffering perhaps most visibly thanks to Artificial Light at Night (ALAN). You can help them out simply by not lighting your garden at night. I've noticed a real trend in domestic gardens of up-lighting trees and hedges, but trees and hedges don't need light at night, they need darkness, and so do the creatures that live in them.

In fact, it's been understood since the 1940s that light exposure can have an extremely detrimental effect on trees. Overexposure to light affects pretty much all elements of a tree's developmental life, affecting leaf shape, colour, root structure and, most crucially, autumn-leaf drop

[§] Insect, not Beyonce.

and spring-bud bursting. Have you ever wondered how trees know that autumn has arrived? Watching the trees fade through that green, summer flush to shades of ochre, umber and vivid reds, before finally the leaves let go of the branches that have held them all summer, drifting to the soils below.

Deciduous trees go through this cycle year after year, but why go to all the effort of growing a leaf only to shed it two seasons later? Broadleaf trees simply cannot gather the water they need to feed their leaves as it gets colder and the water freezes. So instead of trying to keep its leaves nourished, the tree protects itself by dropping all these needy leaves and heads into a dormant period over the colder part of the year. This keeps the tree strong and safe, ensuring its survival into the next year. But how does it know that winter is coming? (I know, Jon Snow.)

Trees are locked into the seasonal cycle of the sun, they know that as the hours of daylight reduce, it's time to start shutting down. It of course makes sense that an organism so intrinsically connected to the sun as a means to photosynthesize, should be aware of how this light (and therefore food source) is diminishing, but it is nonetheless astounding that we don't pay more attention to what the trees know.

As the evenings creep closer, we put lights up outside and under our trees. They know that there is light there, but they can't differentiate between artificial light and natural light, and why would they? Trees have been on Earth 370 million years. Modern human beings have only existed 200,000 years, and artificial lights,120 years. Really, trees have barely had any time at all to even register what an artificial light is as it's existed for 0.000033 % of their entire existence on this planet. But all of a sudden they're under a spotlight, with great beams being installed below their boughs so we can go 'ooh, that's pretty,' and carry on with our lives. The poor trees, however, are having their natural-light cues blocked, leading to them not dropping their leaves in time for winter. This can be catastrophic, leading to trauma, such as weakened roots, limbs, or simply starving themselves over winter as they try to satisfy the hungry cries of the leaves they should have shed still clinging to their fingertips.

There's recorded evidence of this in America, where tree-lined avenues suffered huge losses in storms the very year after artificial street lights were installed, while neighbouring avenues' trees stood strong in inclement weather. Not to mention when we light up our trees, hedges and gardens, we are again removing potential habitat options for wildlife. So, save yourself some pennies and stop it. Please.

We have lit up the lands, but what about the oceans and, actually, the land beneath the sea? Light pollution extends its reaches to the seabed around coastal towns and developments; for example, three quarters of the West Country's

Plymouth Sound & Tamar estuaries are bathed in light, as a team from the University of Plymouth discovered.

Lots of life in the ocean is light shy, and light will actively repel it. Even bright lights on ships are displacing animals. Tiny crustaceans, like copepods, respond to subtle changes in moonlight from their homes eighty metres below the water's surface. That's around twenty-three floors of a residential tower. Imagine how confused they must become by ALAN.

Dr Thomas Davis who has been leading this study at the University of Plymouth and the ALICE (Artificial Light in Coastal Environments) project with Bangor University states

> *Artificial light from coastal towns and cities can now be detected above twenty-two per cent of the world's coasts, nightly. These areas are set to grow even further, as human coastal-population centres are set to double by 2060. We know that many coastal, marine creatures are highly sensitive to light and have adapted to use cycles of moonlight to inform various aspects of their life history. There is great potential for artificial light from our coastal cities, ports, harbours and marinas to interfere with natural light cycles, and the marine animal behaviours that rely on detecting them.*

Most harbours and coastal places around the world flood their seafronts with light, to give people a place to promenade come the evening, or just because it's pleasing to the eye. Bridges over estuaries, straits and rivers are always lit to high heaven, usually iconic structures illuminated for no other reason than to show off. Just look at my home, Ynys Môn, where we have two highly illuminated bridges spanning the Menai Strait to reach the mainland. The light dances on the dark water, twinkling away merrily. It seems harmless, but beneath the surface, these two bridges of light will be acting as barriers, blocking passage through the strait for a whole plethora of species.

Last but not least, our birds. As a nation we are enthusiastic birders, with the RSPB being the largest wildlife conservation charity in all of Europe. My dad is a mad keen birder, and I've inherited the bug. I would dutifully be dragged through waterlogged fields, writing Dad's sightings of curlew, lapwing and greylag geese on scraps of damp paper, kept carefully crumpled in my pockets. (I was much less enthused at the time, with many near misses with bog cotton, and bulls that looked too much like heifers for my liking.)

So how can it be, that this nation of birders is blind to the fact that our light is disrupting our birds in catastrophic ways, and in staggering numbers? In the UK we are lucky to be visited by lots of migratory species, causing sensation across the islands when a bee-eater is spotted over the western shores of Wales, or even a spoonbill in the waters of Norfolk. Our visitors come from all over the globe. Manx shearwater, for example, travel from their nests on the islands off the western edges of Wales to Argentina, a distance of over 7,000 miles. Wales is globally important for these birds, with the world's largest colony living off Pembrokeshire's shores. A brilliant fact from the Ramsey Island RSPB warden, Greg Morgan, is that in its lifetime a Manx will travel to the moon and back ten times, that's nearly five million miles. With an average lifespan of thirty years, that's clocking up an annual mileage of 165,683 miles!

Manx shearwater (*Puffinus puffinus*) are engineered to perfection for life at sea, however, on land, it's a different story. Their legs and feet, powerful for swimming, are far back on their body in order to propel them through the water. This isn't so good for terrestrial travelling, where the Manxies become clown-like as they shuffle along the floor. They're pretty useless quite frankly, which is why they roost on offshore islands safe from mammalian predators like rats and foxes. So dangerously inhibiting is their uncoordinated shuffle, that they nest in underground burrows, only coming into land under the cover of the darkest of nights when the moon is at its newest.

Human encounters with these birds are rare, usually seen at sea in a flock, 'rafting' on the surface waiting for nightfall so they can return to the roost and feed their chicks. It's quite the spectacle to see these amazing birds return to roost. Their call is haunting, so much so that the pirates who once roamed the seas of West Wales mistook it for the cackle of witches (perhaps too much rum and a guilty conscience).

So rare is it that they see humans, it's almost as if they don't know to be fearful of us. Visiting Ynys Enlli in 2020 (between lockdowns), which is home to 30,000 pairs, I came across a number of shearwater as I walked the footpaths at night. As if I wasn't even there, they shuffled past unperturbed by my presence. I had to give one a gentle nudge so I could get by without standing on others. It didn't seem to care, took three steps to the side and looked at me quizzically, its white underbelly illuminated in my torch light.

Untouched by humans as they are, our addiction to light is impacting these birds in terrible ways, unseen by most human eyes. Manx shearwater have evolved to be sensitive to the light of the moon as it waxes and wanes, arriving on land only at new moon in total darkness. When their chicks fledge under the cover of darkness, they do this for a reason. The western horizon glows, naturally, and it has served for thousands of years as a natural beacon to the shearwater. This glow orientates them. Once these chicks take flight they are unlikely to see land again for two whole years, so it's imperative they're not misled by artificial lights from our towns and cities.

Sadly, the bright, city lights are a stronger draw than the horizon's glow, which is growing fainter every year as it's washed away by lighting. This leads to strandings of confused shearwater fledglings, as far inland as Birmingham where they have been reportedly found at the foot of office blocks. Often mistaken for penguins, they're usually picked up, put in a box and delivered back to the west coast of the UK to try again. They can even become grounded on ships, such as the tankers coming into the petro-chemical works of Pembrokeshire. Pembrokeshire Coast National Park actually gives out information to these ships, asking them to dim their lights to stop these strandings, and a wider approach in Wales is being discussed to ask people to be more considerate with their light usage around our coastline.

It's not just Manx shearwater, in fact the majority of our birds are impacted by light pollution. It can come in many forms; street lights, lighthouses, garden lighting and interior lighting, when there's a lot of glass involved. Sat in my friend's house one night, enjoying a nice panad[¶], we heard a huge thud from the French doors leading into the garden. A poor dazed female duck lay on the floor. Thankfully, after a few moments she stood up and took off again, and besides nearly giving us heart attacks, no damage was done. However, this story is repeated the world over and it's not always such a happy ending. Take the US, for example,

[¶] For anyone unsure of what 'a nice panad' is, it's a lovely cup of tea.

where birds and other wildlife are paying a heavy price for our shared urban spaces.

It's been recorded that the 9/11 monument 'Tribute in Light' puts around 160,000 birds at risk every year. If you've not seen the installation, two powerful beams of light are blasted four miles into the sky in memoriam of the tragedy that took place and the many lives that were lost when the Twin Towers came down. In an unfortunate twist of fate, the anniversary of this human tragedy now marks a wildlife tragedy, as it times with one of the largest mass-migration events in the US. Thousands of birds, from geese to hummingbirds, make their migration through what is known as the Atlantic Flyway, a route which crosses straight over the heart of New York City. They become drawn to and disorientated by the lights, leading to them becoming exhausted, injured, and often dying. It also causes the fatal attraction of insect species and bats. Birds of prey, such as peregrines, take advantage of the mass of smaller birds gathering in the sky. The two towers of light have become columns of catastrophe for American wildlife.

The New York advocacy group Audubon works with the memorial organizers, monitoring the impact and counting the birds caught in the beams. Volunteers count the birds, and when the number reaches 1,000, the lights are shut off to allow the birds to disperse before being turned on again. To me it is bizarre that this occurrence continues on an annual basis. Yes, it's important to commemorate such huge human loss, heartache and sorrow, but is this really how we want to hold a memorial? Creating more pain, suffering and loss? There must be another way to harmoniously honour both those that lost their lives and the wildlife that just needs to make its way through.

American cities are also hazardous places for migration due to other sources of light pollution; between 90,000 and 230,000 birds are killed EACH year in New York alone.[5] Open this issue up to the whole of the US and over one billion birds every year are killed by building strike, fatally attracted by artificial lights causing collisions with windows. Birds don't understand glass like humans, it's not perceived as a barrier. To a bird, glass reflecting night sky back at them looks like actual clear sky, so they fly into it at full speed, usually resulting in death. Most species of bird migrate at night, so this is disastrous for huge swathes of the avian world. The NYC Audubon patrol streets in New York during key migration times over spring and autumn, looking for injured and dead birds on the streets. They collect and monitor data, allowing them to pinpoint the city's deadliest buildings and to work with the building owners to make their properties safer for passing birds. In December 2019, NYC Audubon and its partners achieved an important milestone in its campaign, with The New York City Council listening to the group and passing a bird-safe building policy that states that all new construction and major renovations in the city must make use of bird-friendly building design, such as opaque or patterned glass that blurs reflections.[6] This glass was retrofitted into Manhattan's Javits Centre, previously one of the city's worst for bird strikes, which after installation saw strikes drop by an impressive ninety per cent. The new façade also made the building more energy efficient, reducing energy bills by twenty-five per cent, and in turn reducing carbon emissions.

Another charity doing great work around this topic is the Fatal Light Attraction Programme (FLAP) in Canada. They estimate that in Toronto alone there are 950,000 buildings that could kill nine million birds each year. If we took these figures and multiplied them by every city in the world? It doesn't bear thinking about. It is vital that our cities are considerate when designing their sky lines, it's simply unacceptable now to not consider human impact when we build.

This includes our homes, as the *Grand Designs* trend for glass-fronted homes grows, cavernous glass-houses are popping up along our coasts and most beautiful inland areas. It makes sense; glass will allow people to enjoy views, and lots of daylight means no need for lights during the day. However, at night, these homes become giant light bulbs in the landscape. They never seem to have blinds either, which I don't quite understand. Either these homeowners are hugely exhibitionist or they are blissfully unaware they are visible for miles around. These homes become light polluters and potential bird-strike magnets, not forgetting the immediate area around the house also being unnecessarily lit and adding to another loss of habitat. Kevin McCloud – you've a lot to answer for.

I know most people have not thought about the environmental impacts of their lights. I know that most people aren't deliberately using their lights to harass wildlife, but the simple fact is, we can do better. Conservation can be doom and gloom, with seemingly unfixable issues, but I am here to ease your eco-anxiety. What if I told you good lighting could help stop the biodiversity AND the climate crisis?

Here is what you do; you turn your lights off when they're not being used. Simple. You could go and do that right now. You've taken a world-saving step.

Next, when lights are in use, make sure they are directed downward and fully shielded with no light being spilled upwards. That's a waste of your money and is a major contributor to skyglow. Light should only be used to light what you want to be lit, when you want it to be lit. Motion sensors can work wonders. Crucially for our wildlife, the temperature of the light needs to be warm. Kelvin (K) refers to the colour temperature of the light and the colour it emits at 2700K is recommended as best practice in the Dark

Sky world. An ambient light, it's cozier, creating a welcoming atmosphere. Importantly it has less blue-wave content, which is what harms wildlife (and humans).

Once these simple points are covered, the light is tamed and the risk is, if not completely gone, then minimized as much as possible. There are no long-lasting after-effects, no clean-up of light centuries into the future, it is simply gone.

Glazing is a bit trickier, but if you're going to build in a beautiful area with sensitive ecology, it is only right you spend out on making your home have as little impact as possible. Specialist glass that responds to light levels is available, becoming like a one-way mirror as the night draws in, meaning no light spills out, but your view remains unspoiled. If we want to live in these places, we need to *live in* them, not alongside them as separate entities. We are part of this ecological network and any harm we inflict on it, we are also doing to ourselves. Our lives very much depend on us getting these things right. If we don't change our ways, we are quickly approaching a very silent night.

It really is a solution so easy; it can be fixed with a flick of a switch.

CHAPTER SEVEN

Celtic Constellations

To live in Wales is to be conscious

At dusk of the spilled blood
That went into the making of the wild sky
– R.S Thomas

'Dad, I'm *freezing*. I can't feel my nose. Or my toes.' With all the grace of a penguin, I waddled twenty paces behind him, his form briefly encased in an orange spotlight every ten steps or so, before slipping back into the shadows between the lights. My face stung as the salty southwesterlies whipped my skin. I pulled my hood about my face, but the wind refused to let it rest. Battling on, not wanting to be left behind, my frozen fingers wanting to let the hood go but knowing the cold would be much worse if I did. Dad came to a stop just after the last light in the village, on the brow of a small hill, looking into a deep darkness. The dunes were out there somewhere, a lake, too. As my eyes adjusted, I started to make out the jagged line that gave away the presence of Eryri on the horizon, like a horizontal crack in the hemisphere. In the distance the tireless roar of the sea, waves echoing off the hard, frozen shore. 'Come here Dani and look at this.' Dad's face was rosy, the bitter wind causing his eyes to glitter. He bent slightly to reach my eye-line, holding his arm out to the dark night sky. My eyes followed the line, from the end of his fingerless-mittened hand out to a trio of bright stars, hanging serenely over the edge of the world. 'That's Orion.' He traced a shape in the sky, Orion's sword hung from his belt. I squinted, trying to use my imagination as Dad told the story of the mighty hunter. It didn't look much like a man to me, but I couldn't tear my eyes from the sky. He gestured to a bright, yellow star, twinkling in the corner. 'That's Betelgeuse. Orion's shoulder, and this one . . .' he pointed, 'all the way down there is Sirius. His hunting dog.' I was enchanted, the cold instantly forgotten, and spellbound by Sirius, dancing there in the sky. It was like nothing I'd seen before. I wiped the tears from my windswept eyes to make sure its light was really dancing before my eyes. A disco ball in the heavens, flashing red, blue, green. From that night on, I was captivated.

Looking up to the unknown, a huge and beautiful chasm, can be terrifying. Standing under the cloudy core of the Milky Way it's easy to see where stories of vengeful gods and benevolent goddesses started in explanations of something so otherworldly. The night sky is both intrinsically part of our world, yet somehow totally alien. What did our ancestors think when meteors streaked across the sky? It's little wonder that throughout our history, cultures and communities have orbited celestial wisdoms, the stories woven into our very beings and livelihoods. Silver threads delicately connecting the skies to the ground beneath our feet and the water beneath our sails, forever pulling us back towards the darkness.

My memories of learning about constellations will forever tie me to my dad. Frozen fingers pointing out Orion, my first and favourite shape. The great hunter with his star-studded belt, raising his bow and arrow high in the winter sky in an everlasting hunt across the hemisphere. How fortunate was I that Dad knew the stories so well that he could reel them off by heart; I've always joked that his brain was the size of a planet. This is how our star stories have lived alongside us, told down through generations as we gazed up, each telling a little different from the one before. There's a chain extending back through time, an exchange of knowledge, names and folklore reaching back along the starlight.

This age-old exchange is under threat as light pollution looms over the heads of most, blotting out this link to our past. We rarely venture from the warmth of our homes after the sun has set; the flickering light of the TV replacing the flickering lights of the night, opportunities to share our stories fade into the disappearing night, blinded by the lights.

It is for this reason that in Cymru our own star stories are not well known, with many sadly lost as the retellings have slowly stopped. Constellations and their meanings differ around the world, depending on where you are and what culture you call your own. Here in Cymru, we are no different; our lives were once linked closely with the movements in the sky. We make no secret of our agricultural heritage, with thousands of years of farming shaping the landscapes we call home, so it makes sense that we have stories of celestial fields shaped by heavenly hands.

From one of our many hilltop forts, you will see the drystone walls of old sheepfolds and the tumbledown remains of a Hafodty* make constellations all of their own; but once upon a time what was farmed on these green and rolling hillsides would have been helped along by lessons learned from the sky. A time-honoured tradition, before things were written down, was to use star lore to help people remember the wisdoms told before.

* 'Hafodty/Hafotty/Hafoty' – Haf – Summer, Bod – dwelling, Tŷ – house. A dwelling used by farming communities when agriculture was more mobile, in the sense that there were summer and winter pastures, with farmers moving to the mountainside pastures in summer.

Let's look back to Orion, my first constellation, which rises at the start of winter. Before calendars existed, Orion would have warned farmers that winter, and harsher weather, was on its way, signalling it was time to move down to the winter pastures, or the 'Hendre'[†].

In Cymraeg (Welsh), Orion's name is Mabon. He is a deity of the winter sun and announces the arrival of the end of the year's agricultural cycle. In folklore, Mabon was a clever and deft hunter who was the only one strong enough to handle the mighty hunting dog, Drudwyn, called upon by King Arthur to hunt the Twrch Trwyth (an enchanted wild boar). Drudwyn is then represented in Welsh skies as Canis Major, including the sky's brightest star, Sirius, known as the Dog Star in many cultures.

Mabon is also used in the pagan belief system as a name for the celebration of the autumn equinox festival in September, just as Mabon starts to be visible in our early morning skies. There would be feasts held at Mabon – also known as a harvest festival, harvest home, Meán Fómhair (Gaelic, meaning 'middle of the harvest'), or Alban Elfed – giving thanks for the harvest received to get them through the coming winter months.

[†] 'Hendre/f' – Hen – Old – Dref – Farmstead. A dwelling down in the valley used in the harsh winter months.

In some Celtic and Gaelic tellings, Mabon becomes 'Cernunnos', a mythical man adorned with stag antlers and a god of wild places. Sadly, much about Cernunnos has been lost to time.

In winter, the people of Cymru are blessed with many hours of darkness. In the north before the shortest day, the sun sets by 4 p.m., meaning we have a lot of hours of darkness to enjoy! There's been a recent reclamation of our old 'dark' traditions such as Noson Calan Gaeaf (Winter's Eve). Halloween is celebrated in the West on 31 October, but our ancestors would have marked it as the end of the old year and the start of the new. Noson Calan Gaeaf has its roots in the old Celtic celebration of Samhain, where the boundaries between the realm of the mortals and the spirit realm 'Annwn' were at their weakest, and it was easy for the living to slip beyond the veil. Graveyards and crossroads were avoided and feared as places where lost spirits gathered and roamed.

The dark, autumnal nights, the ones with crisp air that smells of woodsmoke and wet grass, are a wonderful time to let yourself ease into the slow season, a time of relaxation and restoration. Make the most of the little light, but revel in the opportunity to envelope yourself in total darkness.

Noson Calan Gaeaf, is often when young children have their first opportunity to be allowed out at night. My memories of Halloween are of bin-bag capes that rustled as we went door to door, the harsh plastic, fake vampire teeth cutting into gums and making speech impossible. Stick on witches' fingers, and food colouring for blood, with cheap, white face paint plastered onto sugar-coated faces. The chaos of multiple siblings trying to carve one pumpkin, mushy pulp stuck between fingers as an endless seed supply is pulled like innards through the top of the pumpkin head. Delight in seeing the candle flicker, illuminating the triangle eyes and teeth, as Mum turned off the inside lights to give our Frankenstein pumpkin full effect.

But before pumpkins were introduced in Wales, we used Rwdans (turnips) that were carved into creepy faces. Men would dress as 'gwrachod' in rags and masks, going door to door for fruit, nuts and other treats. This would scare the evil spirits away – and locals into handing over their treats.

Children were warned to be home before the last embers of the bonfire burnt out, or be taken by Hwch Ddu Gwta (the tailless black sow). Macabre men would don a pigskin and chase the children home from the feast. Don't tell me we don't know how to party in Cymru.

Back to our constellations, and one for spring – the opposite end of the year – when the light is starting to creep back and Earth is on the cusp of rebirth. We look to Gemini, as it's known in the West (in Latin, 'The Twins'). But you may know it as your star sign. The Greek story is widely known and of course involves Zeus being a controlling night-time dick, resulting in a woman bearing twin boys, Castor and Pollux: one a mortal human and one an immortal god. The brothers were extremely close, and together they became heroes, joining Jason and the Argonauts in the quest for the Golden Fleece. Eventually, Castor died in battle and, bereft, Pollux begged Zeus to make him mortal as he could not bear to live without his twin. Zeus took pity on him and immortalized them both in the night sky.

In Cymru, however, the two stars (Pollux and Castor) were not brothers at all, but 'The Rivals of May'[1]. The stars represent two men, Gwyn ap Nudd[‡] and Gwyrthur ap Greidawl: two Knights of the Round Table, duty bound to fight for King Arthur. But instead of fighting for the legend himself, they turned their attentions to each other, fighting bitterly over the love of Creiddylad[§] – the most beautiful woman of the old isles. In one feminist's nightmare of a version, Creiddylad is bequeathed to Gwythyr, infuriating Gwyn, who lies in wait with a small band of men during their wedding ceremony, then dramatically charges down the aisle on horseback, grabbing Creiddylad and stealing her from Gwythyr then taking her to the Underworld. As

‡ God of the dead and the Annwn.

§ Some say she was the inspiration for both Geoffrey of Monmouth's Queen Cordeilla and Shakespeare's Cordelia. Creiddylad is viewed by some as a goddess of fertility, representing the spring.

you can imagine, this leads to an awkward team meeting the following Monday at the Round Table; lots of squabbling ensues. Eventually, Arthur has heard enough. He's being humiliated by his mates down at the local inn as two of his knights are fighting each other instead of getting on with sorting the Saxons out. He slams both fists on the table and banishes them to battle, eternally in the night sky. Whoever he deems the winner on judgement day, gets to marry Creiddylad. Eternity is a long time for poor Creiddylad to wait, so I like to imagine she went off and lived a happy, single life, with no boy drama.

In the tale, and indeed in the night sky, the two knights are locked in an eternal battle. In winter, Gwyn emerges and captures his love. In the sky, the annual fight for Creiddylad commences from November until the month of May. If you watch this constellation throughout the winter, the two main stars, Castor and Pollux, seem to tussle with one another, before disappearing into the summer sun until winter skies return and the battle continues. The story goes that Gwyn (meaning 'white') represents the winter, and Gwythyr (meaning 'victor') represents the summer, his victory marking the light overcoming darkness, the long days of summer returning, and winter being banished, back to the underworld with Gwyn. The victor is celebrated with the feast Noson Calan Mai, Beltane, or May Day. It was then that the agricultural population would start to relocate back to pleasant summer pastures, high on the hill sides of Eryri.

At the foot of the Carneddau, overlooked by lush landscape that sweeps down to the Afon Menai on the outskirts of Bangor, you will find Castell Penrhyn. Built a little over 200 years ago, huge grey slabs of stone create a domineering and formidable stamp on the landscape. It was built by the Pennant family, who became rich from their extensive sugar plantations in Jamaica, and who by 1805 were prolific members of the Transatlantic Slave Trade, owning nearly 1,000 enslaved people[2] across their four plantations.[3]

The Pennant family then funnelled their money into this corner of north-west Wales, opening what was at the time the largest slate quarry in the world, Penrhyn. Still operational today, its lights bleed into the darkness of Eryri. At their height, the family owned 70,000 acres of land, including 700 tenanted farms and nearly 1,000 workers' houses, most of which were built in the town of Bethesda, that sprang into being at the opening of the quarry. At its peak, 3,000 men were employed in the harsh and dangerous working conditions of the quarry.[4] 'The Great Strike of Penrhyn' lasted three years, the longest dispute in British industrial history. The workers and members of the North Wales Quarryman's Union demanded better pay and safer working conditions in November 1900. The strikes were cruel and harsh, many families suffered hunger and poverty, ripping the community apart. The memories live on here and a century later, there are locals who will never set foot in Castell Penrhyn. The walls still cast a shadow over the heart of the community. I only made my first visit to the castle in recent years. I felt traitorous, as if I were stepping over the picket line to betray the people this corner of Cymru was built on.

Now in the care of the National Trust, Castell Penrhyn is a behemoth. The main hall is entered by a faux portcullis, through an impressively carved, stone archway, before the narrow hall opens onto polished marble floors and huge, domed ceilings comparable to a cathedral. No ill-gotten expense was spared by Lord Pennant, it seems, during the build. The hall is intimidatingly huge. My eyes were immediately drawn to two enormous panels of stained glass, in decadent golds, reds and blues. The footsteps of visitors echoed from all sides as I moved closer to study the artwork. To my surprise, I saw the signs of the zodiac on the left panel. They marked out each month of the year with its correlating sign. Somewhere near the top, two prettily patterned fish marked Pisces and the month of March. Roman numerals and the latin 'Lunar' appeared under it, with several circular Celtic knots. The roman numerals seemed to tell the number of days per month. The pieces were so vast, it took a while to fully see them, like going to a cinema and standing right at the front of the screen. I shuffled right as the crowd parted and a man in a short-sleeved shirt and a camera gripped in his hands, asked, 'What's all this about?'

I stepped back to take the full frame in, careful not to be swept into the tide of passing people. The months again were clearly demarcated in calendar order – January through to December. Only this time, instead of zodiac signs it showed agricultural scenes. The confused man had been looking at what he considered a 'bunch of drunk men' but was in fact the month of Leo, representing July. On one side, a proud lion with resplendent mane looks over his shoulder. To his left, the corresponding artwork showed medieval men stooped, not in drink, but in harvest. Collecting the hay into bundles in the summer season, preparing for the winter ahead. The panes displayed an agricultural calendar, as represented by the signs of the zodiac.

I can only assume that the powerful and notable Pennant family had visions of grandeur for this stately home and perhaps after a visit to Paris, once looked upon the stained glass of the Notre Dame. Suspended there, in a building famed for its beautiful stained glass, is an artwork of similar schematics. Known as the 'Labour of the Months', it is an astrological calendar depicting the power of God and his movements of the universe. These were used to inform the uneducated masses how to interpret the actions of the heavens through agricultural labour. We can surmise that this was the message Lord Penrhyn wanted to put out to his tenant farmers through his stained-glass windows. An education on how the heavens manipulated the earth for the peasantry, before reading or writing abilities were widespread. Not that his tenant farmers would have been allowed close enough to see it for themselves.

In medieval times, astrology was central in science, culture and medicine. Like our lives revolve around the sun now, medieval life revolved around the heavens. Miniature depictions of the zodiac were frequently found on calendars, alongside a picture of the Labour of the Month. Sensibly, this does change depending on where you are in Europe, allowing for local climates and practices to be taken into consideration; for example, grapes and wine making featuring in French calendars but not the Welsh. Like little squares from comic strips, these motifs were instantly recognisable and easily interpreted to the peasant classes, a picture literally being worth a thousand words. Some of the calendars are intricate things of beauty, while others are more crude and basic, like a stick man versus a Vermeer. But, be it a window in a castle, cathedral or prayer book, the use of these motifs shows the importance of the night sky and the knowledge it held for people of the medieval age and beyond. You will be familiar with old sayings, such as 'red sky at night, shepherd's delight' meaning a red sky at sunset predicts clear skies and a hopefully nice evening. 'Magic' was often worked into knowledge of the skies and the weather, with 'charms' used by the agricultural workers to try and gain some control over the often-brutal forces of nature. Sayings like these are hangovers from a more superstitious time, but they've carried on being told, because they hold what was once important land knowledge. Planting by moon phase is something still considered seriously across the country, with sayings or incantations such as 'Plant after the first full moon in May' being an echo down the generations, told year after year, to pass on the vital knowledge that before the first full moon of May, any crops planted were susceptible to late spells of frost.

For us in the modern world, living and eating as we do, we are disconnected from the celestial and natural cycles. With the ability to find fresh produce all year round and to eat things like strawberries in deepest December, it can be easy to scoff at superstition. However, these communities were extremely vulnerable to the whims of the weather. One late cold snap, or unseasonal drought, could spell starvation and death for entire villages. We aren't as untouchable as we think we are, either. As farming practices change rapidly, and small farms give way to the larger, more intensive methods, this endless line of traditional knowledge is severed. Thousands of years of hard-earned and learned information about weather; growing, grazing, sowing, the moon and the stars stops with that last voice to tell those tales, and the traditions edge closer to being lost, ploughed back into the fields of time to lie forgotten. Until the system fails, and the modern ways have ignored the warnings in the skies and we find the supermarket shelves are bare, we won't realize until it's too late, that we are still part of this cycle. Most of us can no longer see the cues in the constellations, for light has bleached them away. We think we control our environments, our landscapes and future, but the fates tell us otherwise.

Farmers' almanacs are an attempt to bring all this unwritten lore, knowledge and weather wisdom and place it onto paper. They have existed for hundreds of years, with handwritten versions existing from thousands of years ago. 'Hemerologies' were a version of the early almanac, noting good and bad days to plant crops, get married or even eat fish. Often containing information about astrology, the Babylonians pressed the information into soft clay and left it to harden into a solid tablet. Egyptians, Greeks, Romans, and many other civilizations also had versions of an astronomical almanac.

The first printed almanac to appear in Britain was *The Kalender of the Shepherd*– a translated version of the original 1493 Parisian text, *Le Compost et kalendrier des bergiers*. The first edition appeared in England in 1503, badly translated from French into Scottish (apparently completed by a Scot who wouldn't pass their French GCSE today). English publisher and printer, Richard Pynson, decided to take a stab at it, and his edition was printed in London in1506. With every edition published thereafter, its popularity grew. The book contained everything from astrology and the twelve labours, to religious learnings around the seven deadly sins, and an 'eyewitness' account of hell; all depicted in intricate and sometimes raunchy, woodcut illustrations. With the technological advancements of the printing press, books became much more affordable and accessible to the masses, and almanacs became the best-selling books of the medieval ages, as popular as the Bible, right through the fifteenth century.

A Welsh-language almanac was soon to follow, and in 1679, an eighteen-year-old Thomas Jones of Cae'r Ddol, Corwen, was given a royal patent for the creation and publishing of the first mass printing of a Welsh language almanac ('Yr Almanacc') – *Newyddion Oddiwrth y Ser* – or to those of you unfamiliar with yr hen iaith, *News from the Stars*. The almanacs became hugely popular and were invaluable as a resource to poor farming communities in Wales, containing folklore, moon phases, market dates, tide times and more. Jones even invited amateur writers from Wales to submit articles, an opportunity unavailable to them before. Fishermen were not excluded, with many denoting the best moon phases under which to go fishing and for what type of fish. The National Museum of Wales holds a wonderful archive of many of these almanacs, spanning a good two centuries' worth of astronomical and societal happenings. A number of them have been digitized, so that now, 300 years later, you can read news of the stars from the 1680s.[5]

Almanacs provide a fascinating peek into our heritage, and each edition plots the gathering of knowledge through time; they truly are a treasure trove of social and folk history, and their popularity continues to this day. A 1556 edition of *Shepherd's Kalendar* sold for £13,145 at Christie's in 2002.[6] The importance of these almanacs shows us just how significant astronomy and the art of watching the night sky was. Whole lives and livelihoods in Cymru depended on an ability to read the sky. We are intrinsically connected to the darkness, not just in Cymru, but in the British Isles and beyond. That was, until light pollution came along. Stealing this knowledge from our collective memory. Can you imagine what an incredible sight unpolluted night skies were to those generations before us? To step out into a dark night and see thousands upon thousands of stars, out into the universe, our home. Being unable to see the stars means this ancient wisdom is passed on no longer. With each flickering bulb, we lose a little bit more of our connection to the cosmos, and along with it, our culture and heritage.

The stars don't just hold knowledge of the land, but the seas and rivers, too. For generations, Cymru has been shaped by the sea around her shores and the veins of waterways that trickle, stream and pour from mountains to marshlands. How did we sail the Seven Seas before satellites could show us the way?

In a world of Google Maps and satnav, it never fails to surprise me that satellite GPS only came in to use in the late 1970s. Up until that point, the world's shipping trade operated by celestial navigation, compass, and physical maps.

At the Porthmadog Maritime Museum, I stand in a darkened room, surrounded by nautical paraphernalia that adorns every inch of wall, ceiling and floor. The dark timbers wedged in the pitched roof give you the feeling that you're trapped under an upturned hull of a wooden boat. I am drawn to a glass case that is softly illuminated from within, glowing among the rows of wooden ships, and a female figure who once adorned a ship's prow stares serenely into the distance to my side, her black hair swept up behind her, reminiscent of a Snow White of the seas. In the centre of the glass box, sits a beautiful brass instrument. Upright, it resembles the shape of a capital 'A' atop a rocker like a rocking horse. It looks to me like an alien relic. I know that it is a sextant, but I could tell you no more, I assume it must be hundreds of years old. Next to it, folded open, is a small and worn notebook. The pages are well-leafed. My nose is almost to the glass as I squint to read its handwritten notes.

Before I fog up the glass, an older gentleman appears by my side. He's a volunteer here. 'Would you like a closer look?' he says, and clicks open the case, selecting from a ring containing many different keys, the right key, first time. He retrieves and hands me the notebook. 'Do I not need gloves or something?' I ask, before taking what I expect to be some fragile piece of history. He laughs. 'Gloves? It's only my notebook!'

I am immediately confused. 'Your notebook?'

'Yes,' comes the reply. 'And that's my sextant.'

How is a man old enough to have used a sextant at sea standing here in a museum next to me? I place him as in his early sixties, latest. I look at the notes written on the page open in my hands and see the date written – '1st March 1968'. We switch to Welsh language and he talks me through his notes, written on yellowing paper with blue grid-lines showing me how they marked their location on the globe using the stars' positions at night. Weather, speed, observations, all are noted down with a complex series of numbers and equations. In my naivety, I imagined sea navigation by starlight to be a thing disappeared from our zeitgeist at least a century ago.

'We could have sailed straight out from Porthmadog, turned left and set sail for Brazil,' he tells me, and I am in awe as he shows me the thousands of miles of journeys he navigated as a merchant sailor. They would sell their goods in Brazil and pick up new goods, like crates of oranges, sugar and meat before setting off again. He said they could call at ports across the world, all along the eastern coast of the USA to Canada, back to the Caribbean and across to Portugal and home again.

'Could you teach me how to use a sextant?' I ask. He frowns a bit, then tells me that sadly, in 1975, the GPS came in, and overnight everyone stopped using the old methods.

'I couldn't remember how to do it now,' he says. 'It's quite a shame really. I worked this way for years, we all did. But you'd be hard-pressed now to find someone who can use it, and really use it well.'

He was right. I've tried for years to find someone to show me the ways, and many people know the basics, but no one confident enough in their ability to teach it. Interestingly, this loss of knowledge is raising concerns further afield. The US Navy are now trying to bring the old navigational system into use. As satellite dependency increases, we leave ourselves vulnerable to attacks. The US has already experienced 'satellite jamming' from hostile countries and expects this type of space warfare to increase.[7] In the case of a rival shooting down a GPS satellite, they would be literally adrift, with no way to navigate. Teaching new recruits the 'old ways' of navigation is a sensible back-up plan in the face of new threats, I believe.

Without understanding the night sky and knowing its movements intimately, we would never have set sail from the sandy shores. We would have no trade or communication links across the Celtic Sea, nor had visitors, friendly and hostile, from lands far, far away, who influenced our culture and language. We would not have had the infamous Barti Ddu and Sir Henry Morgan, pirates who set sail from Pembrokeshire, or Madam Wen from Sir Fôn. Without the stars, Swansea would not have become 'Copperopolis', and the slates from the heart of Eryri could not have 'roofed the world'.

Another darkness lies at the heart of Welsh history – 'black gold' coal. Mined from the valleys of South and West Wales, many people spent their lives in darkness, underground. Their labour shipped around the world from half of the Welsh population would not exist if it wasn't for the ports that brought newcomers from over fifty countries to start new lives in Cymru's docklands.

Coracle men on the rivers of the Teifi and the Towy used their traditional star lore to know when the perfect time was to push their coracles gently from the riverbanks and float serenely in search of sewin, salmon and trout. They would wait for the 'seven stars' to be visible in the sky, fishing only in the hours of 'cyflychwr' (dusk, or twilight), between the 'dau olau' (the two lights).[8] The crepuscular coracle men wove their crafts with superstition and folklore, raising their nets to the first winds of the season and weaving the hair of their children into the ropes of the nets for luck. By waiting for the seven stars, they would have enough light to fish by, but not so much light that they would cast shadows in the water and scare the fish away. The coracle community ebbed and flowed with the river, respecting nature and the animals they shared it with. The curlew, the 'new moon' bird was held in high regard, judging the coracler's moral compass and keeping them safe from harm. Similar to Viking tradition, when a coracle man died, the custom was to burn his nets and his craft on the banks of the river, but only once they heard the call of the owl, believing that this wise old bird would guide the departed soul up to heaven. Birds, stars and folklore have long nested together in the branches of Welsh history.

I love to walk outside in the hours of cyflychwr, when day is slipping into night. That brief period that lingers after the sun has dropped below the horizon; not light, but not yet dark, either. In late autumn and winter it's the time you can see huge congregations of all sorts of birds. A walk to Coed Niwbwrch (Newborough Forest) and you will hear the raucous ravens, roosting in the towering Corsican pines, their roots enmeshed with the dunes, the anchor to this privateering crew's home in the canopy. Their conspiring carries through the dense needles and cones.

From my lowly position on the woodland floor, a 'conspiracy' of ravens, as they're collectively known, certainly sound like they are concocting a clever scheme or two. It's an odd place, where the treeline meets the shore. A plantation created to protect the residents of nearby Niwbwrch from inundation by sand, completely artificial yet home now to over two thousand ravens, living their secret lives up in the trees. Hidden from me where I am on the ground, is the life of this court; the rituals and displays bonding each pair for life; the emerald mosses that line their treasure troves of precious eggs in shades of speckled sapphire and aquamarine. The blues of their nests are traditionally symbolic of wisdom, intelligence and loyalty, reflected in the nature of these characterful corvids, the largest of their species.

As your feet find the edge of the forest floor, it gives way to dunes and a flat expanse of golden sands. At low tide the sea leaves an architectural maze, a miniature of contours, canyons and capillaries. Water left behind, trapped by ridged dams of the exposed seabed, creates pools of mercury, polished and placid, shining the lingering light of dusk back into the sky. A rush of air overhead announces the steady stream of shapeshifters, starlings (from the Old English for, 'little star'). They arrive in small numbers at first. I watch them appear in the sky, one by one until the sky is studded in their tiny black silhouettes. They make their way in a long, oscillating line, covering miles in the sky, reaching out like a river winds across the ground. No murmuration, no sky dances, but they seem to draw the darkness in with them, as they head around the coast and inland over the village of Aberffraw.

Hovering over Ynys Llanddwyn, Jupiter arrives. Fitting that the god who had a dubious record with women would appear above the island of Santes Dwynwen, the Welsh patron saint of lovers, who fled to this tiny island in the fifth century to live out her life as a nun, due to men being, well, men. From Ynys Llanddwyn you see seabirds gathering for the night on the small outcrop of rock called Ynys Yr Adar ('Island of the Birds'). Here, a huddle of comical cormorants, prehistoric-looking birds, stretch their wings out in the fading rays, drying their glossy black feathers as if suspended on a crucifix or outstretched in offering, worshipping the sun, asking it to return to them again as their world is swallowed whole by the night. Their Welsh name changes, depending where you are. 'Bilidowcar' is what I call them, but in other places they're called a morfran ('sea crow/raven'), a name shared with our Cornish cousins. From their perch they look west, towards Ireland. Cailleach dhubh ('the black hag') is the name they hold there. Dad always taught me they foretold stormy weather, and that finding a cormorant inland meant high seas and winds were on their way. Tonight, they have no need to worry. The ink-pot sea turned darker still, until a small, silver glow of the faint waxing, crescent moon, joined Jupiter in the sky. Together they set sparkles into the sea, light bouncing on the smallest of ripples on this cold, clear night.

In times gone by, birds held significance in Welsh culture. It is of no surprise that they make their mark in the heavens. From January till May, a celestial corvid migrates to roost in the night skies of the north. 'Corvus' ('the crow/corvid') is a personal favourite as it connects some of my favourite Welsh myths, birds, and places together, weaving them into a story in the stars. It starts in the Royal court of Aberffraw, a few miles down the road from where I grew up. In school, my 'house' was that of Branwen ('White Raven'). It is she who this story concerns, and her (literal) big brother, Bendigeidfran, who happened to be a king and a giant. He also goes by the name of Brân Fendigaidd ('Blessed Bran/Blessed Crow/Raven'). It's a long and complex story, and in true, Welsh folkloric tradition, it is not one that ends well for the women involved. In a much-butchered version for you to enjoy, Branwen is married in Aberffraw to King Matholwch of Ireland. However, due to a despicable happening on the night of her wedding in which the King's horses are maimed by her petty half-brother, her life in Ireland is not a happy one and she is forced to work day and night as a scullery maid in the depths of the castle, enduring violence at the hands of men, and the other servants are ordered not to show her an ounce of kindness.

One day, a starling with a broken wing comes to the window and a lonely Branwen befriends it, feeding it bread from the kitchens and nursing it back to full health. She cares for the starling, day after day, and teaches it to talk. She tells it of her brother, the giant Bendigeidfran and explains how to find him. In real life, starlings are great mimics. They have been known to mimic the noises of children playing, telephones ringing and dogs barking. When fully recovered, the starling flies across the Celtic Sea to Cymru, finding Bendigeidfran and delivering the message. On hearing how his sister is suffering, Bendigeidfran raises his army, and a huge fleet of ships sets sail for Ireland, with all the warriors of Wales.

Bendigeidfran, too huge for any ship, wades alongside them, just tall enough to keep his mouth above water. An epic battle ensues – and believe me, the whole story is gorier than George R. R. Martin's infamous Red Wedding, which is thought to be based on this tale. The whole of Ireland is slaughtered, save for five pregnant women. Only seven Welsh warriors survive. Bendigeidfran is mortally wounded by a poisoned arrow. Knowing he is dying, Brân commands the survivors to remove his head from his body as the Celts believed the head was the home of the soul. Our poor Branwen, bereft with grief at the destruction she has caused, dies of a broken heart when she reaches the shores of Ynys Môn. Upon landing at Aber Alaw she cries, '*Oi, a fab Duw! Gwae fi o'm genedigaeth. Da o ddwy ynys a ddiffeithwyd o'm hachos i! (*'Oh, son of God, woe to me that I was born! Two fair islands have been laid waste because of me!').

Branwen's burial site remains there on the banks of the Alaw to this day. A small cairn where her earthly remains were laid to rest. As for Brân, his head is taken to London and buried at White Hill (on which the Tower of London is now thought to stand) facing towards France as to protect the lands of ancient Britain from attack. The Celts believed that when people died, their souls became or were conveyed to the otherworld by birds, and so Brân's soul left his body and became the mighty Raven, or 'Corvus', in the night sky. As both Brân and Branwen's name means raven, I like to associate them both with this constellation. In the story, Branwen shows courage and patience, she is as much a hero as her brother. I feel it's only right to have her represented in the cosmos, too.

There are ravens at the Tower of London to this day. Legend tells that if the ravens abandon their post at the Tower, Britain will fall and the Crown will crumble. In 1675, Britain's first Astronomer Royal set up his telescope at the Tower. John Flamstead, appointed by King Charles II, was tasked with creating an accurate map of the night sky that could be used for navigation. Flamstead quickly became irritated by the ravens of the tower, who flew across his lens, obscuring his view and sometimes even fouling the telescopes. Flamstead persevered with his cosmical challenge despite the ravens, before moving into the newly commissioned Royal Observatory in Greenwich and even asked the King to get rid of the ravens.

An injured raven was found on the dunes of Aberffraw in 2013 by Mr Unite.[9] He called the Ravenmaster at the Tower of London for advice, to find that they actually had space for a new raven. According to tradition, six ravens must be kept at the Tower at all times, with another two as back up, or I suppose, spares to the Heir. The lucky raven was perfect for the job; having an injured wing meant she couldn't desert her post. *Blue Peter* ran a competition to name the new resident raven soon afterwards.

In 2021, disaster struck when Merlina, the matriarch raven at the Tower died. Known as the undisputed 'Queen of the Tower'[10], her death was quite the omen. It wasn't long before a new chick was found to take her place within the royal roost. Another vote was held, with members of the public voting for their favourite name. What were the replacement ravens of 2013 and 2021 both called? Branwen, of course.

In Cymru, we really love giants, and we like to name places after them. Our place names can also reflect a counterpart in the night sky, the mountainous area that includes Cadair Idris[¶] for example, just outside of Dolgellau. Gawr Idris was a Giant King, who was so large he used the mountain as a huge armchair, Llyn y Gadair (also known

[¶] 'Chair of Idris'

as 'The Seat'), wrapped within the arms of the mountain ridge. He was a warrior poet and astronomer, who would use his mountainous perch to spend hours gazing up into the heavens and over his kingdom. There's a saying that any person who was to spend a night on Cadair Idris would rise the next morning to find themselves 'a mad man or a poet'. These hillsides are also known to be the hunting grounds of Gwyn Ap Nudd and his ethereal Wild Hunt, rounding up lost souls to take back to the Annwn (the 'underworld'). If you were to take your chances with losing your soul to the King of the Tylwyth Teg ('fair folk') you would be rewarded with some of the darkest skies Eryri has to offer. Maybe they've been protected from light pollution by Idris Gawr himself, as even now the stars still appear in their thousands. If you're lucky, you might just make out his form in the constellation known more commonly as 'Hercules'.

I am lucky enough to hear these stories brought to life regularly by wonderful storytellers who are masters of their craft. One in particular, Fiona Collins, is a fantastic chronicler. I am grateful to her for imparting to me many of these wonderful tales, and for always finding more!

As I finish this chapter, Jac a'i Wagon** are rising high into the night sky. A constellation most are familiar with, it has innumerable names. Commonly 'The Plough' or the

** Jack and His Wagon.

American 'Big Dipper', one Welsh name sees it as a traveller's wagon being pulled across the sky. This large shape is an 'asterism'[††], part of a huge constellation, and Ursa Major, ('the Great Bear'), the third largest constellation in our night sky. It has been used for navigation throughout the ages, thanks to its ability to guide you to the pole star and guide weary travellers on their way north, just like the sailors who used the night sky to navigate the seas. It is fitting that it is the friend of travellers, for it is these people who brought their stories to our shores, sharing them with us, mixing them with our own and creating new ones along the way. These stories have travelled through time and their reflections can be found in the stories of other cultures from around the world; the night sky connecting the constellations of communities throughout the world and history. Weaving our national identity on a loom filled with threads spun in places far from our own, combining to create the magnificent tapestry that is Cymru today.

[††] Asterism – A recognized pattern of stars that is not part of the 'official' list of eighty-eight constellations.

CHAPTER EIGHT

Under One Sky

The sky, our common and universal heritage, is an integral part of the environment perceived by humanity. Humankind has always observed the sky either to interpret it or to understand the physical laws that govern the universe.

– Proclamation of 2009 as International Year of Astronomy, UNESCO, Paris 2005

For many of us it may seem that the world has never felt so divided. Every day we open our eyes and scan social media and news outlets, reading never-ending stories of bad news. Things seem to be happening at an almighty pace, like we are ricocheting, faster and faster, into an uncontrollable storm. Wars, poverty, climate change and crisis after crisis. Right vs Left, 'War on Woke', it can appear that Britain is truly 'broke'. It's overwhelming, and that's without personal concerns and worries.

Now that you're thoroughly depressed, I urge you to go out and look up. There is nothing like standing in the universe, under the immensity of the night sky, to be reminded about how out of our control all of this really is. The enormity of the great beyond never ceases to put my worries into perspective. Taking in the sights and sounds of the night, with thousands of stars suspended in the darkness overhead is incredibly grounding. Sure, things might not be great right now, but time marches on regardless. The universe continues, and so will you. Feeling insignificant has never felt so good.

Never in my lifetime have I felt more of an urgency to force humans around the world to look for our common ground; strengthen our similarities instead of searching for differences to exploit and turn us against each other. There is nothing more universally binding than our shared heritage in the night sky.

Every single culture across the globe, throughout our existence, has looked up and tried to make sense of it all in the stars. It is the view to the night sky that has made us question where we are from, who we are and where we will go. This desire to look up is something so primitively human, and it has driven humankind to create some of the most significant and beautiful things in the known universe. From millennia-old cave paintings to Van Gogh's *Starry Night*; Josef Strauss's *Spharen-Klange* (*Music of the*

Spheres) to Bowie's 'Starman'; from folklore to the rocket science that put us on the moon, there is nothing else that unites us and drives us as much as a star-filled night sky.

Yet it is being stolen from us. Our access removed through light pollution and satellites, with very few remaining places on Earth where you can see a truly Dark Sky. Still, there are fewer places again that are doing anything to protect their skies and control light pollution, erasing before our eyes and at the speed of light, a right to access our most undervalued resource; one that is rich in global heritage and cultural significance. By removing this link to ourselves, we loosen the knot on our tie to the natural world, forgetting our place within it and our responsibility as stewards of this planet: to leave this world a better place than when we got here.

Looking at the star lore of cultures around the world, you will start to see the similarities and common themes that draw us together. These stories have drifted across oceans and down rivers in canoes, boats and coracles. They have been carefully carried over mountain ranges, deserts and plains, to be shared at firesides and campsites, passed from ear to ear as surely as loaves passed from hand to hand. Each of the cultures' star lore mentioned in this chapter is worthy of its own book, too complex and nuanced to do justice to in a single chapter. Instead, I aim to give you a taste of a small selection, a starter. At the end of this book, you will find links to lots of wonderful resources should you wish to delve further into the global heritage of our night sky around the world.

The great Pimoe

We will start in the lands farthest from my own – Polynesia, some 7,000 miles from where I sit now. Polynesia is a constellation of islands, covering a huge area of the Pacific Ocean, from the Hawaiian Islands in the north to New Zealand in the south, with Tonga, Samoa and Easter Island, to name just a few of the islands, in between. Each island has its own cultural fingerprint, and while they share many similar cultural traits, each has its own distinct societal and political structure. One thing that unites these islands is their history and reputation as some of the globe's earliest and most accomplished mariners.

Their seafaring journey as a people is testament to their knowledge of the natural world and the night sky, using the stars to navigate to some of the most remote and uninhabited places on Earth. Sailing from Asia, they headed eastward in the search of new homelands sometime around 2000 BC*. Setting sail in their canoes, east by starlight until settling on Tonga and Samoa as some of the first islands;

* Roughly during the same time as Britain's Bronze Age.

hopscotching their way across the vast expanse of ocean until eventually founding populations right across this huge area of the Pacific, covering a span of thousands of years and thousands of miles. A phenomenal feat of human adventure and seafaring ability. Their way-finding technique relied on an encyclopaedic knowledge and excellent observational skills that were passed down often in the form of song. To navigate by the night sky, a 'star compass' was used, but unlike what we think of today as a compass, the star compass was held in the mind. A catalogue of constellations and stars whose patterns and movements were known to the voyagers and used to navigate successfully. On cloudy nights, other natural clues were used, such as wind direction, wave patterns and the movements of fish and birds.

Clearly, the stars held significant knowledge, and the history and heritage of the Polynesian people only exists because of their relationship with the night sky. Without it, they would not have become the successful navigators so venerated today. So of course they have a wonderful and rich selection of star lore to choose from.

One story is about a giant, ulua fish called Pimoe. It is king of fish, residing deep in the ocean, prized by fisherman as it is so difficult to catch. One day, three brothers set out by canoe into the beautiful blue waters of the Pacific in search of Pimoe. One of the three was Maui, a demigod. He is determined to make the catch and land the legendary fish. After a time, his brothers start to make fun that he has caught nothing, let alone a prize Pimoe. Suddenly, a flash of silver is seen and a splash that sets the canoe rocking causes the three to steady themselves on the wooden sides of their hollowed-out craft, their knuckles turning white with effort. 'Pimoe!' Maui cries, casting his fishing line out into the direction of the tail splash. His eyes are on the prize; the hook at the end of the line glistens in the sun before plunging down, deep into the sea. 'Paddle, and do not stop!' Maui instructs his brothers as he feels the line go taut. It stretches and strains, and his two brothers paddle furiously, their shoulders burning and their muscles screaming, as a large and shadowy shape starts to loom beneath the canoe. 'Do not look back, keep paddling!' The shape looms larger and larger as it gets closer to the surface, tantalisingly close to breaking the barrier between the realms of water and air. Maui's muscles bulge with the effort as he uses all his strength to land his glorious catch. With one last mighty yank he pulls the shape free of the sea. There is rumbling, water sloshing, waves rolling towards distant shores. The canoe rocks violently as the brothers' eyes close with the force of paddling and hauling in the catch. When they open them again, they are shocked to see the canoe now sitting on a black sand beach, with a luscious landscape behind them, where the open ocean had been before. Instead of a Pimoe, they realize Maui's hook must have snared the seabed. Maui had pulled up a new island. His brothers are amazed, but Maui, furious that Pimoe had escaped, flicks his hook free of the newly created shore and sends it skyward, where it remains to this day as the constellation of 'Manaiakalani', the name of Maui's magical fishhook.[1]

This is said to be how the island of Hawaii came to be. In alternative tellings, it is New Zealand, or other islands. Who knows, maybe Maui got a taste for heaving new lands from the seabed and created many of the islands that stand proud in the water today.

The night sky and navigational links to Polynesian culture run deep in the veins of its people, but even they came close to losing these ancient ways of wayfaring. Luckily, realizing this pillar of their identity was close to being lost, efforts have been made to record the knowledge, and islanders taught the ancient craft of sailing by stars, saving it from 'cultural extinction'.[2]

They have fought hard to bring this knowledge back. The Polynesian Voyaging Society (PVS) are quite simply, incredible. In the 1970s the Hawaiians were facing this cultural extinction. There had been no traditional canoes for 600 years, the knowledge was all but gone. Determined to bring back this part of their identity, a *wa'a kaulua*[†] was built

[†] A traditional double-hulled Polynesian canoe.

with the intention to bring the ancient crafts back. *Hōkū le'a*'[‡] was launched in 1975. The story is heartbreakingly heroic. The aim was to sail to Tahiti from Hawaii and back using no modern navigational tools. But no one was left who knew how to navigate using the old ways. The PVS managed to recruit a Master Navigator, Mau Piailug, from the solitary island of Satawal in Micronesia. A tiny community of just 500 people, the island was the last bastion of these old navigation techniques. He was among just six people left who knew them. He was also the only one willing to share his knowledge. Without Mau, these traditions would almost certainly have been lost. The path to resurrection has been wrought with tragic losses, including crew member and big-wave surfer, Eddie Aikau. But in 1980, under the watchful and knowledgeable eye of Mau, Nainoa Thompson became the first native Hawaiian to sail to Tahiti by natural navigation methods in nearly a thousand years.

The PVS now teaches these skills, restoring cultural pride among its people, curating this intrinsic link between Pacific Ocean islanders and the natural world; one that Western colonialism has done much to sever. Now they sail thousands of miles, visiting conservation projects and educating communities on how to take care of the planet.

[‡] Meaning 'Star of Gladness', the Hawaiian name for Arcturus.

To think that these losses could happen again, but this time through needless light pollution, is shameful. It is nothing but cultural erasure.

The night sky from down under

Now to the oldest continuing culture on the planet, First Australians. For over 60,000 years they have resided in Australia and could well be the world's oldest astronomers. Their songs, art and stories have been passed through thousands of years and contained within them is sophisticated astronomical knowledge. It is important to note that there are hundreds of Aboriginal nations, including Torres Strait Islanders, each one unique and therefore with its own constellation names and stories. As astronomers, they are distinctive in the way they incorporate the whole night sky and not just the stars, but also where there are *no* stars, known as 'dark constellations'. They take into consideration the darkness between the starlight. Perhaps the most famous of these is *Gawarrgay,*[3] *(*the 'Emu in the Sky'). When you look up at the Milky Way, you are looking into one of the arms of our home galaxy. At certain times of the year in the UK, we can see what looks like a dark river, running straight through the centre of the Milky Way's arch above our heads. That is known as the 'dark rift'. It is essentially clouds of gas and dust, that block our view to the stars, making it look like a great ravine has been gouged out of the Galaxy. In the Southern Hemisphere, it looks like a great celestial emu. As our position through the year changes, so does the emu's. These changes are important to Aboriginal culture, dictating the timing of essential ceremonies such as 'Bora' (Initiation), the availability of food, seasonal changes and also informing people of the life cycle and behaviour of emus, such as breeding and egg-laying seasons.

This indigenous knowledge was recently celebrated in 2020, when two beautiful limited-edition one dollar coins were commissioned by the Royal Australian Mint as part of their 'Star Dreaming' series. The first, designed by Scott Towney, a member of the Wiradujuri tribe, was silver, embedded with a blue, star-studded sky and, of course, the noble emu, pride of place in the middle of the sky. The second coin was created by a Wajarri-Noonjar artist, Christine 'Jugarnu' Collard, and is an equally beautiful depiction of the story of the 'Seven Sisters', or the *Pleiades*. An Aboriginal dot painting depicts the seven stars in vivid blues over a red landscape, then cast in silver are seven women, running from the unwanted advances of a male suitor. These coins are commemorative of the ancient knowledge, showing respect to the indigenous people of Australia, but is this same respect shown through protections of the night sky?

The Wardaman people of Australia's Northern Territory, believe that the darkness in the Milky Way is where the Spiritual Ancestors reside, as told in the book *Dark Sparklers* by Bill Yidumduma Harney and Dr Hugh Cairns about the homes of the Ancestors.

The Ancestors created all of life on Earth. They gave knowledge to the humans on how to protect this earth, told them of the sacred places and how to honour them, and they taught people how to treat one another and showed them the laws of the land. When their work was complete, the main Ancestors returned to their homes in the sky. It is in the dark spaces of the Milky Way that these spirits reside, each night they watch over the people and make sure that they care for Earth and each other.

The Dreaming is a period of time that had a beginning but no visible end, when the natural environment was shaped and given humanity by mythical beings. To the Aboriginal people, the Dreaming is the past, the present and the future and they are aware of it in everyday life, it is central to their world view and culture.

Very simply, Aboriginal communities value the night sky as it helps explain their world. The ethos that runs through their stories, both in the sky and beyond, show the significance of stewardship and the importance of place. It is something the rest of the planet should be listening to, extremely carefully.

A lot of their culture was lost when colonizers settled in the 1780s. Many settlers exploited this knowledge, (alongside all the other awful crimes committed against indigenous peoples) especially when it came to travelling and wayfinding. Songlines are pathways laid down across the landscape, navigational routes memorized by song and passed on. There are examples of constellations being used as navigational wayfinders, pathways over the landscape that were along Songlines. An article by Robert S. Fuller from the University of New South Wales, suggests that if you overlay the wayfinding constellation with a modern road map of Australia, they match perfectly, showing how accurate their knowledge of the night sky was.[4] Now Dark Skies are under threat from colonialism in space, through satellites, rocket launches, and light pollution meaning this culture will lose even more than it already has to the West.

China's Qixi Festival

One of my favourite stories from another culture is from China. So significant is the story that it is marked with a national holiday on the seventh day of the seventh month of the Chinese lunar calendar, called *Qixi Festival*, which is comparable to St Valentine's Day.

The story starts with the Jade Emperor, ruler of heaven, as he creates the sky. It is long, difficult work, creating the sky, so he asks his seven daughters to help. The seventh princess of heaven is Zinhü, a talented weaver. Her father asks her to weave clouds of beautiful colours and mists as fine as silk.

After working for hours, she becomes tired and asks if she may take a break among the newly created stars. The Emperor agrees, but Zinhü and her sisters decide to leave heaven and visit Earth instead. There, they bathe in a river, and a young cowherd named Niulang stumbles upon them while driving his herd. Shocked to see seven beautiful maidens in the waters, he hides himself away. There he spots their robes on the banks. He cannot take his eyes off Zinhü, falling immediately head-over-heels in love with her. He hides the princesses' robes, and when they emerge from the river, they are annoyed to find their clothing missing. Niulang steps forward from his hiding place, eyes lowered, blushing furiously. He knows that this is no way to introduce himself. He tells the women they can have back their robes, only if Zinhü agrees to marry him. Her sisters demand that Zinhü take him up on his offer, so she agrees. Luckily for Niulang, Zinhü immediately fell in love with him the moment she saw him and agrees to remain on Earth with him.

For a while they live a happy life and have two children, but before long, the heavenly Queen Mother hears of this marriage and is enraged that one of her daughters has married a lowly mortal, and not even a rich one. The Queen orders Zinhü be captured and returned to heaven. The two lovers are heartbroken. So moving was their love for each other, that a god spoke to the heavenly Queen and asked her to reconsider. It didn't end well, with the god being banished to Earth as an elderly cow, who then goes to Niulang and tells him that when he dies, Niulang must use his cowhide to cover him, and then he will be able to travel to heaven. With that, the cow dies and Niulang does as he was told. Wrapping the hide around himself and his two children, he finds himself able to visit the sky to see his beloved.

When the Queen discovers this, she becomes further outraged and removes her golden hair pin, striking it across the night sky, where it leaves a river of stars in its wake – the Milky Way. So vast was this river of stars that the lovers can not reach each other. Their tears fall down to Earth and their sorrow is heard and felt throughout the land, by all the creatures of Earth. But it is the magpies who fly up into the sky, an uncountable flock, and use their vast numbers to span the Milky Way. Their wings and feathers interlock to form a bridge for the two lovers to meet upon, the iridescence of their magpie feathers sparkling in the light of the heavens.

So beautiful is the sight, even the Queen cannot deny the two any longer. She agrees that Niulang and Zinhü can meet on the magpie bridge every seventh day of the seventh month. In the night sky you will see their story play out, as on the seventh of the seventh, the stars of Altair (symbolizing Niulang) and Vega (symbolizing Zinhü) come very close to the Milky Way. Bridging the river of stars is our Magpie Bridge, represented by the constellation of 'Cygnus' (Greek for 'swan'). This constellation is made of six bright stars that make a cross shape in the sky, 'bridging' the gulf of the Milky Way for our two lovers to meet on that one day a year.

This story again shows the strong link between birds and stars in folklore. There are many variations of this myth throughout Asia, including in Korea, Japan and India, but this telling goes way back – 2,600 years at least – first appearing in China's *The Classic of Poetry,* traditionally said to have been compiled by Confucius.

Many cultures link birds and stars, such as the Finnish, who call the Milky Way *Linnunrata* ('the Pathway of the Birds'), after recognizing that this pathway in the stars was followed by birds when migrating, sending them south for winter. We now understand that birds really do use celestial navigation when migrating, and the removal of stars through light pollution causes migratory birds to become disorientated and lost.[5]

The Milky Way, when viewed from a place with little light pollution is awe inspiring. It is difficult to comprehend that we are looking out, into our own home galaxy, within which we drift among thousands upon thousands of stars, neighbours to our own magnificent sun. To the generations before light pollution, the view of the Milky Way would have been simply breathtaking. To those who didn't yet have the science to understand what they were seeing, it is little wonder that there's a wealth of mythology hung up there on that awesome arch, as we have tried to make sense of our lives under it.

South American nights

To the Tukano people of Colombia and Brazil, the universe is quite literally at the heart of the home. Their 'malocas'[§] are built to be reflections of the cosmos here on Earth. They are big houses, designed meticulously to enable the community to live under the roof in harmony, not just with each other, but with the natural world.[6] The malocas are built east to west, following the direction of the sun in the sky.[7] The impressive palm-woven roof is held up internally by large posts, but above them they are joined together by one central beam known as the *gumu*. This represents the Milky Way. To the Tukano, the Milky Way connects the three worlds: the underworld, Earth and the upper-world.

Even the constellations bear extreme importance to survival. Each animal that is hunted for food by the Tukano has an association in the stars, but importantly, that specific animal can only be hunted for food after their constellation

[§] A type of longhouse.

has risen above the horizon.[8] In their mythology, the animals are aware of this, and hide when their constellations appear. Through systems such as this, the Tukano are ensuring they do not over-hunt or take more than is needed. This self-regulation shows a deep understanding and respect for the world around them and how they are a part of it, something I wish I could transplant into the brains of Western civilizations as we continually take up more resources than we need, obliterating the natural world, instead of nurturing it. The very existence of these people is intrinsically woven into their responsibilities and stewardship of the natural world upon which we all rely. Light pollution would without doubt create deep distress and turmoil in these communities, ultimately degrading their culture.

The Milky Way is often seen as a road or bridge that connects realms and worlds. To the Native American Apache people, this stream in the sky is a trail made by departing spirits, it rests on the shoulders of Yolkai Nalin, the terrifying Goddess of Death. All souls have to pass her in order to make it to the afterlife. The Navajo people have a story in which after the First Woman created the sun, she went on to carefully carve the moon from a glittering white, quartz rock. As she carved, pieces flew from the moon and created the stars in the sky.

In Spain, the Milky Way is referred to as *El Camino de Santiago*. Originating in medieval times, the Milky Way was believed to have been created by the dust from the feet of thousands of pilgrims, and the path of the galaxy in the night sky seems to follow the famous pilgrimage trail, the 'Way of St James', which ends with travellers planting their feet in the celestially named, *Santiago de Compostella* – which translates to 'Field of Stars'– the saint's resting place.

A fervour for the Aurora

Heading north, the Scandinavian countries are famed globally for their colourful night skies that dance with myth and legend. These stories tie their cultural identity to the darkness that envelopes these northernmost lands. The closer you travel to the Arctic Circle, the longer and darker the night stretches on. In places like Lapland in Finland, the sun hangs heavy in the winter sky, barely able to keep its head above the ground until it can hold on no longer. When it finally gives in and bows below the horizon in late November, its rays will not lighten the sky or warm your skin again until mid-January. Even for the darkness-lovers among us, this can be tricky to live with, for humans need a balance of light and dark. I can't imagine saying goodbye to daylight for winter. Countries nearer the Arctic Circle, with vast periods of night-time, are overflowing with folklore about darkness and stars. For these countries, it is also a huge tourism draw.

Increasingly, people across the world want to experience darkness and an endless night, and the opportunity to see not just a star-filled night sky, which many will never have experienced but, of course, the Aurora Borealis. The vivid light shows that grace Northern European skies. Mysterious and beautiful they reduce many to tears (I confess – I cried like a baby). Before we understood about the sun's coronal mass ejections[¶] (not a euphemism), it was hard to make sense of what was happening in the depths of winter, when on the deepest, darkest nights the skies suddenly exploded with riotous dancing colours. They appear all at once, without warning, like ribbons in the sky, in hues of pink, red, green, purple and blue, forming all manner of shapes and spirals. The ones I saw formed pillars at first, gently toppling into one another, like luminous sky dominoes, before swirling into a whirlpool of astonishingly bright light, while I stood rooted, in awe, my feet freezing in the Icelandic snow.

Perhaps the most famous of stories connected to the Aurora Borealis is that of the Norse god, Odin and his legendary cohort of Valkyries. Beautiful war maidens, sometimes winged, sometimes on horse- or wolfback, they were feared and revered. It was believed that the Aurora

[¶] Coronal mass ejections are solar flares, that shoot energy out from the sun creating a solar wind. This mixes with Earth's magnetic field and creates the Aurora.

foretold the coming of the Valkyries, the lights reflecting off their armour, spears and shields as they rode over battlefields, selecting only the most noble of slain fighters to be taken with them to the hall of Valhalla, and join Odin in a glorious afterlife to await Ragnarök. Another bridge across the night sky into the otherworld.

For the indigenous Sámi people, the lights are to be fearfully respected. They hold the belief that if the lights become aware of your presence, they could come and spirit you away, up into the sky. So many Sámi stay indoors. The lights are the souls of the dead, and making any noise under them is thought to be disrespectful. They even remove the bells from their reindeer, so my sobbing would have been very much unwelcome. I fully understand this fear. Lucky enough to witness an extremely powerful aurora in Iceland, I was not prepared to feel the way I did, they came so low overhead and seemed so electric I could feel myself cowering, shrinking into the snow. It was quite frightening. What really threw me was the *sound*. I had no idea you could hear the lights. It was almost like a faraway audience in applause, hundreds of softly clapping hands. That night I think I prayed to almost any god going if it meant my feet remained rooted to the snowy ground.

In Iceland, it is said that if a woman is in childbirth the lights will ease the pain. Whereas in Greenland, almost the opposite is thought, with the lights poignantly having the connection to souls of babies who didn't survive birth. Inuit people also believed them to be representative of their dead, the lights a way to communicate with their deceased.

A lesser-known story is from the Hebrides, an archipelago off the west coast of Scotland. There, the lights are known as *Na Fir-Chlis* – which translates to 'The Nimble Men' or 'The Merry Men'. This name is somewhat misleading as it belies the story that the lights are a violent fight between warriors in the sky or fallen angels. So violent were the battles that blood would rain from the sky, staining the rocks and creating 'bloodstones', a form of green jasper, spotted with flecks of red. It isn't hard to figure out which visitors to these Scottish islands brought their folklore with them and left it behind.

My favourite tale, perhaps, belongs to the Finnish, who believe the Aurora is a mythical fox-like creature called Tulikettu (in Finnish mythology, a fox with 'fiery fur'). The fox remains hidden by day, its fur as black as night, but come dusk it starts to twinkle like the embers that escape the flames of a fire. Hunters would pursue this fox, chasing it through the night. It would run across the land whipping sparks of glistening ice and snow into the air. The Finnish name for the Aurora Borealis is Revontulet, meaning 'fox fire', so if you're ever lucky enough to witness the Aurora, look for the bushy tail of a fox across the forest tops.**

** Less romantically, it's also where we get the Firefox web explorer name from.

Despite so much tourism and culture being carved out from their night skies, it is surprising that these northern nations – except Denmark, which had two of its small islands, Møn and Nyord, designated as Dark Sky Communities in 2017 – have taken no steps to protect their night sky through joining the International Dark Sky Association. You would think that to have something so special, they would want to protect it by simply controlling lighting. I was shocked when I visited Reykjavik, the Icelandic capital, that the city had no lighting controls and was quite the light polluting city. Although Reykjavik Council has been known to turn street lights off when displays of aurora are overhead, they could and should be doing much more to protect their skies. I hope that countries like Iceland will soon follow in the fight against light pollution, especially as they have so much to lose. Although, one enterprising Icelander, Einar Benediktsson, did once try to sell the Northern Lights to Swiss businessmen. He was, of course, unsuccessful. I would imagine plans were scuppered when the issue of how to transport the lights from Iceland to Switzerland were discussed.

During my visit, I was alarmed to find out that a light-polluting art installation by Yoko Ono, *Imagine Peace Tower*, had been placed on Viðey island, just off Reykjavik, in 2007. I'm sure Yoko has done lots of good in the world, but this is no way to commemorate her husband, John. It's a collection of fifteen searchlights that burn as brightly as a giant death ray nearly three miles into the sky, illuminated each October through to March – two peak migration times for the world's birds. Ironically, Yoko Ono has been quoted as saying, 'Iceland is a magical and beautiful country. The electrical energy source for the country and the *Imagine Peace Tower* is geothermal … No pollution. No war.'[9] No pollution? Light pollution doesn't count, it seems.

It's not just in the skies that we have placed these stories. When we look at the human record and the collection of art, music, science and literature, we can see that it is quite literally, star studded.

Night-time art

From the cave paintings in Lascaux, where human hands pushed pigment into the rockface depicting a celestial event, to more modern classics like Van Gogh's *Starry Night*, it is undeniable that the stars have played the role of muse to many artists over the centuries. *Starry Night* is a painting most of us will be familiar with. I had to recreate it at primary school, along with *Sunflowers*. I can still smell the oil pastels and remember the satisfying weight of the slickening blue sheen as I layered blues and yellows to replicate what is one of the world's most famous paintings of a swirling blue, night sky, adorned with a glowing crescent moon and stars above a sleeping rural scene, a small village with a towering church spire and the rolling hills beyond. I don't think my version even made the fridge, but this painting is one that has stood the test of time since its creation, so familiar and comforting is its scene.

It was in 1889, the very year that the world's first street came to illumination through electricity, that Van Gogh sat before his easel and captured the magic of a night sky on canvas. As a child in class, I found his paint strokes captivating, so swirly and transient, I felt like they were moving on the paper. Even seen on a grainy copy, you can make out the textures. I wanted to trace the colours and clouds, circle the stars with my fingertips. As an adult I value the painting even more. The clouds that lie low on the hills look like they are moving across the scene, the stars glowing as they do of an evening, when the atmosphere around us is turbulent. I appreciate that Van Gogh was captivated by the Morning Star. Venus sits almost centre, glowing brighter than the stars around it. The scene is one Van Gogh would have seen from the barred window of his room, a glimpse of the night just before daybreak from the Saint-Paul Asylum his brother had placed him in following his breakdown, when he cut off a sizeable part of his left ear. Van Gogh had struggled with severe mental-health issues most of his life. As an outlet he turned to nature and had a lifelong appreciation for the countryside. I can only imagine how beautiful a sky he saw, and how much solace that must have brought to him at a painful point in his life. Many people quote the first part of this letter from Van Gogh to his brother, but the whole paragraph is much more telling of his mental state.

> *But the sight of the stars always makes me dream ... Why, I say to myself, should the spots of light in the firmament be less accessible to us than the black spots on the map of France? Just as we take the train to go to Tarascon or Rouen, we take death to go to a star.*

Van Gogh died just a year after he completed *Starry Night.*

The previous year he'd painted *Starry Night Over the Rhône*, which showed gas lamps marking the River Rhône as it flows through the French town of Arles. Painted in his signature style, he used shades of blues and yellows, blended to make a vivid impression of a star-filled sky, their light reflected and merging with that of the gas lamps on the banks. The Great Bear constellation (also known as Ursa Major) takes centre stage, burning brighter than the street lights and arching across the sky, under which lovers gaze upwards. In a letter to his brother, Theo, Van Gogh describes the painting and says that he painted it at night,

'... actually under a gas jet ... On the aquamarine field of the sky the Great Bear is a sparkling green and pink, whose discreet paleness contrasts with the brutal gold of the gas.'

It seems Van Gogh wasn't overly keen on street lighting himself, even the dim gas lamps that would have been a new addition at the time, viewed as impinging on the natural beauty of a star-filled sky.

Sadly, if you were to visit the location of either painting today, be that Arles or Saint-Remy-de-Provence, you would no longer see a night sky full of stars. Covered in a smog of light pollution, the constellations have all but been banished to reside in memory, captured in time a hundred years ago at the hands of Van Gogh. As a legacy to this artist and his life, the areas should, in my opinion, do everything they can to protect that most famous scene that has captured hearts and minds for over a century. Somewhere, the next great *Starry Night* sits unpainted, invisible to the eye and mind of the artist who could create it if only they, too, could see the sky that Van Gogh did. The night sky has so much merit as a place of historical importance. Imagine we simply left our buildings and other landmarks of historical importance to the same fate? There would be an outcry. But the night sky is being smothered so silently and suddenly that barely a soul has taken the time to notice.

From 'Twinkle, Twinkle Little Star' to Beethoven's 'Moonlight Sonata' all the way through time to Bowie's

'Starman' and Coldplay's 'Music of the Spheres', the stars have always sparkled with musical inspiration. There are quite literally thousands of songs about stars, space, planets and the moon, but there was a point in time when humans believed there was literally music in the stars. 'The Music of the Spheres' was, pre-Coldplay, theorized by Pythagoras (yes, of the triangle), who theorized that Earth was encapsulated by eight transparent spheres, each one holding the sun, moon, planets and stars around Earth in their orbits. Remember, this was when people understood Earth as the centre of the universe. On their orbit, they created vibrations, and just like a string vibrating on a violin creates a note, so did these orbits. It was believed for hundreds of years that there was heavenly music being made up there in the spheres beyond our ears, unless you were Pythagoras, who believed he was the only human spiritually enlightened enough to hear the beautiful harmony. With the planets being massive, the 'harmony' they produced should have been absolutely deafening, but that was theorized away by the belief that as it was there from birth, it simply disappeared into the background noise that only the very enlightened could 'tune' into, which is very convenient.

'There is geometry in the humming of the strings. There is music in the spacing of the spheres . . .' – a quote attached to Pythagoras; from the sixth century BC all the way to the

Renaissance of the sixteenth century, this theory was popular with scientists, but the concept is such a beautiful one that it still influences musicians today. I don't know what it was supposed to sound like, but I'm sure planetary music would sound incredible.[††]

Without our views to the cosmos, we would not have had such poetic prose, verse and plays, with writers captivated by the beauty of a wild and dark night.

Shakespeare's Romeo and Juliet would not be 'star-crossed lovers', nor would Juliet have uttered the following lines:

Take him and cut him out in little stars,
And he will make the face of heaven so fine.
That all the world will be in love with night.

They just don't write pick-up lines like they used to.

The human story is written in the stars. If we carry on polluting our skies, the cultural cost to communities around the world will be yet another tragic loss on our watch. Once these old traditions, tales and skills are lost, it is nigh on impossible to bring them back. Generational gaps too yawning to bridge, even with all Earth's magpies.

[††] Unless it's supposed to sound like jazz.

It's a huge price to pay for the sake of being careless with lighting. The solutions are frustratingly simple to implement, and the benefits far outweigh any perceivable negative. We are on the cusp of it being too late to do anything about this. Already we have lost too much. Let's not lose any more.

CHAPTER NINE

The Final Frontier

It is late. So late that the midsummer sky blooms indigo. The sun, barely below the horizon, prepares for its ascent into the colours of dawn. I've been driving for eternity when finally, the sleeping streets melt into open fields and roads shift from sheets of tarmac to narrow, winding ribbons. Street lights speckle the rear-view mirror as I leave them behind me, the engine biting into a steep climb. I've reached the heathered edge of the North York Moors. Gorse, bracken, and bramble grasp at the road's edge, threatening to ensnare it and drag it back to the Moors. My headlights sweep around corners, illuminating the land to my left, which rolls away into a steep drop. No light escapes from this ravine, a black hole drawing my attention from the road, threatening to suck everything in like a drain.

I refocus on my mission, keeping my eyes peeled for a sign I'm sure must be around the next bend, then I see it: 'Use dipped beam when approaching guardroom'. I cautiously trundle on, looking for the turn. Suddenly, an apparition; thirty pairs of eyes glow silver in the beams, a paranormal flock of satanic-looking sheep. Lucifer leads them, the Morning Star visible but sinking fast into the fiery furnace that is the beginning of another day. A hulking tetrahedron lurks into view, dominating the horizon, a black behemoth absorbing all light around it. I spot the turn and dip my headlights. My shoulders ache, brain buzzing from over caffeination, but relief floods my bones. I know there's a bed for me soon. Struggling without full headlights and a brain full of fog, I fail to see the stopping point, triggering a powerful spotlight that pierces the road and my retinas. Chain-link fences spring up around me, and black and yellow barbed-wire blocks – like brutalist bumble bees with razor sharp stings – barricade the promised land of sleep. I slam on the brakes. Two camouflaged legs march into the light, eventually revealing they belong to a suspicious and weapon-wielding man in military greens. 'GET OUT OF YOUR VEHICLE!' he shouts. *Shit*, I think, desperately searching for my ID, emptying my bag on the front seat. The light seems to intensify, the glare blinding, hampering my ability to think. I slowly open the door, trying to be as unassuming as possible. *Do I put my hands on my head?*

'Can you please help me? I'm supposed to be at RAF Fylingdales.'

Silence. I've lost all sight of the military man with a gun. I feel the caffeine pinging round my brain and squint into the interrogating light. He emerges from the shadows, content that a five-foot-one woman wearing a skirt and heels is not quite the terrorist threat he was expecting. 'Who's sent you here?' he asks.

I am eventually deemed risk free and waved through the barriers. I just want to climb into an uncomfortable, military-issue bed and get some sleep before the birds start.

My experience with the RAF has always been this; you could have told everyone on the entire station the exact time and place you will be arriving, but they will still be surprised when you appear at the gate and a huge scene will unfold. I enter the eerily silent accommodation block and find the right door, comforted to see my name scrawled on the name holder. Military accommodation can be worse than student halls. Unsettling, fluorescent lights reluctantly clunk into action, with the same identikit room at every base in varying degrees of magnolia and blue. A smaller than single bed, a sink furnished with a tarnished mirror, a sign reminding you to run the tap for two minutes lest you get Legionnaires and die, and, of course, the all-important ironing board. I collapse on the bed and check my phone. The screen shows multiple notifications, but my access is now blocked by the total lack of signal. Communications are a no-go. This is RAF Fylingdales. The UK's worst kept, top-secret military base.

Built in 1962, Fylingdales,* was to be a top-secret facility. However, being positioned in the beautiful North York Moors National Park, high on the skyline near Whitby, while also having some interesting architecture, the base soon became a tourist hotspot. Originally home to three 130-foot geodesic domes, these were state of the art radar and highly sophisticated bits of kit. True to form, the great British public soon christened the cutting-edge technology, 'The Golfballs', immortalizing them in the landscape. They became so iconic that there was public outcry and petitions started when, at the end of their life, the MOD tore them down.

A sinister looking sub-woofer now stands in their place. Technically speaking, it is a collection of radars mounted to a looming sarcophagus known as 'The Pyramid'.

RAF Fylingdales' Latin motto is also rather sinister – '*Vigilamus*', meaning 'We are watching'; and watching they truly are. This station is so secretive because it serves a hugely important role, in not just international relations, but intergalactic ones too. It scans the skies, looking for potential threats; ballistic missiles from hostile countries,

* Disappointingly not called Flyingdales.

and danger from outer space. From the launch of the Soviet Sputnik 1 in 1957 right up to this very moment in time, Fylingdales has been keeping track of everything ever sent into space.

If you were around during the Cold War, you've probably heard of the infamous phrase 'Four-minute warning.'[†] It's this station that would deliver that warning. A mountain of pressure rests on the shoulders of the personnel here who have little under two minutes to react to information on their screens. They must determine if it is hostile or an error, and inform the next level that could lead to a counter strike. Get this wrong or fail to act quickly enough, and their actions could have devastating consequences not only on the lives of those in the UK, but for everyone living on the planet.

Simplified, it is an early trigger warning that raises the alarm in the US that a missile is headed in their direction.

The RAF have almost always played a noisy role in my life. Growing up in a village at the end of a runway, I am accustomed to fighter jets, helicopters and other aircraft flying at breakneck speeds overhead. At school, classes would frequently be paused while jets took off and landed, several times a day. Tables shook, windows rattled, and the teacher would sigh, hand on hip as she waited for the roar

[†] Not the 2003 Mark Owen song.

to dissipate. These days, Zoom calls are made quite interesting when you're drowned out by the roar of a Typhoon that decided to stop by for a visit.

There's an old and very true adage about pilots – 'How do you know a fighter pilot is a fighter pilot? They'll tell you.' I swore to never go near a military man, with most of the military partners I knew struggling to balance family life, constant uncertainty, and the impossibility of holding down employment with a mobile lifestyle. So, you can imagine my absolute dismay, upon moving to South Wales and starting a job after university, I met a tall, dark, and handsome man who told me he would soon be packing his bags to join the RAF. I couldn't believe my luck. I told myself that that was the end of that, packed him on his way and thought I'd never see him again. Dear reader, I married him.

Since then, he's been stationed in all corners of the UK and even been flung as far as the Islas Malvinas (Falkland Islands). Most weekends I would travel to various stations, and during his officer training at RAF College Cranwell, I visited the battered and exhausted trainee as much as I could.

Approaching the college by road, you pass grand, gold ceremonial gates. The heraldic emblem of the RAF is mounted on the metalwork, an eagle encompassed in a circlet with a crown. Around this circlet, the motto *Per Ardua*

Ad Astra.[‡] Although not their intention, the RAF have located almost all their stations in Dark Sky areas. From the North York Moors and Ynys Môn, up to Boulmer on the north-east coast and Spadeadam on the edges of Kielder Forest Dark Sky Park. Unfortunately for these places, the RAF are also usually monumental light polluters. Any chance of being covert under cover of darkness completely blown by the clouds reflecting the glow of thousands of light bulbs back to the ground. They certainly are bringing adversity to the stars.

Wandering around the officers' mess at Fylingdales, I found a board of staff names and photos, recognizing one face. I knew this person was once stationed with my partner, but didn't understand the string of letters next to his name. The military absolutely lives for an acronym. Most of them take so long to explain you may as well have just said the words in the first place. After successfully connecting to the Wi-Fi, I sent my husband a message asking what the mysterious letters were short for. What he sent back blew my mind – 'Space Command'. The UK, it transpires, has its own Space Command, complete with the Latin motto *Ad Stellas Usque,* roughly meaning 'Up to the stars'.

Space technically starts at an altitude of sixty-two miles above sea level, an imaginary line known as the 'Karman

[‡] For the non-Latin literate: 'Through Adversity to the Stars'.

Line'. If you took Hadrian's Wall and stuck to it vertically, you could walk into space before you got to the end of its seventy-five-mile length (you'd also be extremely cold and a little bit dead, so it's not advisable).

To get there, you head vertically into the sky, making your way through the many 'spheres' of Earth, traversing the clouds that shroud the troposphere, before stopping to take in the sights of our stratosphere. Here you would need your Factor 50, as this is where the ozone layer lives, protecting us down here from the sun's UV rays. At 29,000 feet, you can just about enter the stratosphere if you conquer Chomolungma (Everest). At this height, you may be surprised to see a whooper swan or crane fly by, taking advantage of the more stable air conditions to aid their migrations.

Reaching the heady heights of the mesosphere, you will truly start to freeze to death as you experience the coldest temperatures in the Earth's atmosphere. At minus-101 degrees Celsius, not even Wim Hoff can save you. If you were suited and booted appropriately, this level of the atmosphere has some wild phenomena for you to enjoy, such as noctilucent clouds: beautiful, shimmering threads of clouds visible after sunset for only a few weeks either side of the summer solstice from Earth. If you're lucky, you might have some close encounters with bits of space junk and space rocks as they hit this layer and burn up, becoming beautiful molten streaks or shooting stars, taking your darkest, deepest wishes to their fiery graves.

As your frozen corpse breaks free from the mesosphere at mile sixty-two, welcome to space, Earthling, you're officially an astronaut.

An extremely useful area exists in space at an altitude between ninety-nine and 1,200 miles above Earth's surface. This little band around our home planet is the sweet spot for human space-activity. It's not so low that your multi-billion-pound science project gets pulled back into Earth's atmosphere, erupting into a fireball as it plummets to the ground, but it is just low enough for humans to use for a multitude of reasons. This band is known as the Lower Earth Orbit (LEO). Modern life is increasingly reliant on satellites and the race is on for companies and organizations to claim their 'orbital space'; their thread around the planet to which they can send their satellites. Humans have essentially created a satellite motorway, busier than the M60 ring road on a Friday night, and getting busier all the time. The satellites have a huge variety of uses, including mobile-phone coverage, the internet, weather, and navigation. If you have used Google Maps to navigate your way home without the stars, checked what the weather is doing tomorrow, or paid by bank card to purchase fuel, you've used satellites.

Satellites also gather imagery and data for things like climate change, ocean temperatures, changes to the polar regions and wildfires. At the other end of the spectrum, they're used for finding valuable resources like minerals and oil. And of course, most importantly, we can all look at our houses on Google Earth and see if any of the neighbours have been caught in their dressing gown in the garden as the street-view car drove by.

Interest in the LEO has led to the dawn of a new space race. Lots of companies are becoming space prospectors, but instead of gold, it's satellites they're after.

Satellites have dramatically changed and improved our lives. However, they have their flaws. We are very susceptible to damages and loss of connection due to space weather. Pretty as the Northern Lights are, if you're in the south of the UK and the Aurora is dancing over your house, you might want to put your electronics in a lead box, deep underground. For an aurora that strong to appear, we would be getting hammered by some powerful electromagnetic matter, fired from the sun. The Aurora Borealis can interfere with satellites and radio waves, leading to a loss of mobile phone and internet coverage, navigation systems and even ground flights and damage the electrical grid.

This orbit is also home to one of the greatest examples of human collaboration in history, the International Space Station.

In November 1998, the first module of the ISS was launched into orbit, a historically significant moment in which some of the world's most conflicting powers put aside their political differences to create a peaceful collaboration in the name of science.[§] A Soviet module was connected to a US module and just like that, the two become one.

As we were still celebrating surviving the Y2K Millennium Bug and Kylie Minogue had us spinning around, the year 2000 saw the launch not only of the first *Big Brother* hopefuls wanting to launch themselves to stardom, but the launch of three brave souls starwards – NASA astronaut, Bill Shepherd, and cosmonauts, Yuri Gidzenko and Sergei Krikalev. The second their airlock opened and they floated through the hatch, they set the clock ticking on what has now been a decades-long continuation of human habitation in space. We may not be zipping around in flying cars (yet) but we have this testament to human spirit and endeavour pass over our heads sixteen times a day. Most of us here on Earth never even notice this beacon of ingenuity as it bursts across our night sky, reflecting the light of the sun down to Earth before fading back into the shadow of our own planet. Now, that must be some sunset they're seeing, sixteen times a day.

[§] Of course, Geri had also left the Spice Girls this year but, as we've learned, not all hope for humanity was lost.

So far, satellites sound pretty good, don't they? Everyone is playing together nicely and then you dig a little deeper.

> *America has always been greatest when we dared to be great [. . .] We can follow our dreams to distant stars, living and working in space for peaceful, economic, and scientific gain. Tonight, I am directing NASA to develop a permanently manned space station.*

These were the words of President Ronald Reagan as he directed NASA to build the ISS. Sounds great, doesn't it? Well, then we get into the real crux of this expansion into space:

> *Just as the oceans opened up a new world for clipper ships and Yankee traders, space holds enormous potential for commerce today. Companies interested in putting payloads into space must have ready access to private-sector launch services ... We'll soon implement a number of executive initiatives, develop proposals to ease regulatory constraints, and, with NASA's help, promote private sector investment in space.*

That's right. Under the pretence of global cooperation, peace and bettering life on Earth, we get to the capitalist core. The privatization of space. I still maintain that the ISS is one of our greatest human achievements and a demonstration that borders are stupid, and science is something that should always overarch politics, but that science is increasingly being funnelled into the possibilities of commercial gain.

From our formative forays into space to today, most of what has been sent into orbit remains. We created a new issue; space junk. This is particularly an issue in the LEO where most satellites are sent.

Unfortunately for the personnel working at Fylingdales, space junk bears an uncanny resemblance to a nuclear war head on a radar screen. In the name of not accidentally starting a third World War, it is imperative they keep a log of the locations and positions of all space junk, satellites and meteorites hurtling around Earth, and keep an eye on those looking like they might head right for us. They have an impressive catalogue, being updated constantly. At last count they were helping to track 2,000 functional satellites and over 30,000 pieces of space debris, from the Tesla car with its mannequin driver 'Starman' at the wheel that was launched into space for no good reason, down to things like pliers, a spatula and a tool bag that have been dropped by astronauts on space walks. These are the pieces big enough to track. There's another 500,000 pieces of rubbish, bobbing around like plastic flotsam in our oceans, that can't be detected from Earth.

It's a growing issue. Even the smallest bits of rubbish can cause catastrophic damage, depending on what they hit and when. Things in orbit are travelling over 22,000 mph and can make holes in very expensive bits of equipment like satellites, telescopes and the ISS itself.

Humans often do stupid things. I mean *unfathomably idiotic* things, like launch satellites at great resource and cost, only to blow them into smithereens a few days later. It isn't rocket science to know that blowing up stuff in Earth's orbit isn't the best idea, but apparently, even rocket scientists allow this to happen. A number of nations suspect each other of developing anti-satellite weaponry (known as ASATs) so blame each other for the reason they've developed their own. An ASAT is essentially a missile that can be launched from Earth to completely obliterate a satellite.

Some of the planet's most intelligent brains have done this on several occasions. The US, China, India and, most recently, Russia, have all blown up satellites, seemingly just to flex their galactic muscles in the direction of other nations, demonstrating capability of destroying the satellites we are now very much reliant on for everyday life. It is reckless behaviour, creating thousands more pieces of space debris and putting lives at risk. China destroyed their own defunct satellite back in 2007. Officially, they removed a dead satellite from orbit. From a military perspective, this was posturing. China has just proved their ASAT capabilities, arguably, kicking off Space Race 2.0; and this time it's a lot more serious for every one of us.

The resulting creation of potentially hazardous debris caused international outcry, especially from the US military, realizing they were slipping behind in another race for space.

In February of 2008, Russia drafted a treaty looking to prevent 'an arms race in outer space', which was backed by China. Weeks later, the US launched their own ASAT, on the grounds that a defunct satellite was heading back to Earth, posing a risk to American lives. So, they blew it up. This was not looked upon fondly by the other nations. Long story short, there have been some sneaky space operations since then.

In 2019, the then President Trump funded the first new US military service in over seventy years; the United States Space Force (USSF), currently the world's only independent military branch dedicated to space defence. Although in its infancy, it already consists of over 8,000 military personnel (the entire RAF has about 30,000) and seventy-seven spacecraft:

> *Space is the world's newest war-fighting domain, amid grave threats to our national security, American superiority in space is absolutely vital. And we're leading, but we're not leading by enough. But very shortly we'll be leading by a lot.*

A truly Trumpesque quote from the former president shows there's no doubt about it, space is a new theatre of war and establishing Space Force was confirmation that the US intend to dominate this final frontier. Unfortunately for Trump, Space Force was widely mocked; from valid points questioning why it needed green camouflage uniforms when there are no trees in space, to genuine concern from defence specialists and other states. Although easy to laugh at, this is a very real threat that future generations will be navigating, with technology advancing at breakneck speeds. It is vital that nations keep working together, to ensure space is kept peaceful.

On 15 November 2021, because the world wasn't feral enough at this point, Russia decided to take out a two-ton satellite located in a very vulnerable section of the LEO, shattering it into 2,000-plus pieces of trackable debris (and many smaller, non-trackable pieces). For the seven, soundly sleeping, astronauts and cosmonauts on the ISS, a terrifying wake-up call came from NASA. Mission Control directed them to get to the spacecraft docked at the space station – the escape spacecraft – confirming that an uncontrollable cloud of lethal space debris was heading straight for the ISS.

This is the stuff of nightmares. You're in a spacecraft, you look out and see the beautiful blue hue of our planet below, but beyond that, there is nothing but a black void.

There is no emergency cord to pull. No quick swerving to one side. This is the pinnacle of a worst-case scenario. Russia knew this. They are a hugely accomplished nation when it comes to space. There is no way they didn't realize the chaos they were unleashing, nor that they were jeopardizing seven human lives. Most astounding, they had two of their own citizens on the ISS and carried on regardless. Mother Russia was not looking too maternal to the cosmonauts aboard the ISS that day.

As scientists on the ground realized, the debris cloud would endanger the ISS every ninety minutes as their orbits intersected and the crew prepared for evacuation. The first two hours must have been a petrifying wait, praying the worst wouldn't happen. By some hand of fate, they evaded catastrophe and eventually were able to return to their abnormal normality aboard the floating, space laboratory. However, this cloud will continue to spread and threaten other objects in space, remaining in orbit for decades to come. Russia is not alone in its actions, and other nations have been making shows of force that come dangerously close to the weaponization of space.

Possibilities of mineral extraction from celestial bodies such as the moon, asteroids, and other planets, also exist, just like invasions and explorations here on Earth can have a foot in economic gain, so can space exploration.

There are already a number of 'space treaties' aimed at keeping peace in space including the Moon Agreement from the UN Office for Outer Space Affairs, which states:

> *Those bodies should be used exclusively for peaceful purposes, their environments should not be disrupted, the United Nations should be informed of the location and purpose of any station established on those bodies.*

Only eighteen states have signed up to this treaty. Peace in space, currently does, and always will, depend on the goodwill and morals of nations around the world to do the right thing.

It's concerning that exploitation of resources on the moon and elsewhere, isn't a universal no. For an intelligent species, we have such little capacity to learn from our mistakes when it comes to the economic gain of people in power over global good. We are talking about mining. Extraction. Removal and disruption of elements we have absolutely no human right to. As we hollow out the planet at huge cost to the climate and life on Earth, with zero indication of stopping, many are setting their sights out into space.

Another night during Lockdown; another clear night. The skies have just tipped into total darkness and the temperature plummets. I am taking my allotted exercise time as an evening stroll around the lake that I was extremely privileged to have on my doorstep throughout the many national lockdowns. The first few stars are pinging into view, as if someone is flicking on the switch to each individual star. I stop to do up my coat, as cold air snakes its way through the threads to my skin, sending a shiver down my spine and transfiguring my breath into smoky tendrils. I follow its spiralling trail to the sky and notice something peculiar: a very bright satellite is heading right for me from over the sea, easterly along the arm of the Llŷn Peninsula, seemingly almost level with the mountains that are dissolving into darkness. It's moving eerily slowly. I adjust my glasses as directly behind it I see another, and then another. A perfectly orderly line of countless lights stretching out into the distance. An involuntary chill ripples down my back as I try and comprehend what I'm seeing. The legion of lights marches by my eyes, travelling over the domain of the Roman Empire's most western reaches. I realize this is a new empire. The Elon Empire.

This was my first time witnessing what is known as the 'string of pearls' – a train of newly launched satellites from SpaceX: the US aerospace company run by South African-born tycoon, Elon Musk. Since 2019, sixty satellites every fortnight were planned to have been sent up into Earth's LEO. These launches, known as Starlink, will continue to go on until the mission objective is complete; a mega constellation of over 40,000 satellites completely covering the planet. The aim? Bringing free internet to the masses. A noble and needed cause across the globe. Many communities and countries find themselves left behind through a disparity of access to the internet. Every person deserves a fair bite at the apple, but as our lives are increasingly moved online, some can't even see the apple as everything migrates behind screens. It's something I have been impacted by; living rurally on a Welsh island, access to the internet can be woeful. Access to a phone signal even worse. Internet infrastructure can be pricey, especially digging up tarmac to lay fibre-optic cables; but not impossible. It can make a monumental difference to people's lives, especially the isolated, giving easier access to services like healthcare and education remotely.

However, Starlink *isn't* free, as originally claimed. It's expensive and getting even more so. It works by satellite dishes on the ground picking up internet from the satellites in space. So far it is priced out for a lot of people in the UK, let alone for people living in less economically developed countries, and communities where access to clean water is a much more pressing priority. At the time of writing this, the monetary cost to access Starlink internet is £90 a month, on top of a set-up fee of £493[1], with costs seemingly going up, not down, the more satellites are launched. Not exactly affordable, and definitely not for those the company originally claimed to be helping.

Venezuela regularly tops the charts for having the world's worst internet access, but with an average monthly wage of just $25, realistically this isn't helping anyone who needs it. What this translates to is further widening of the access gap. Countries that struggle with social, political, and economic turmoil could really benefit, with communications infrastructure, that's non-existent in many places. However, the wealthy will be the ones who get to benefit first, and arguably the most. As for those of us with access already, how fast does our internet need to be? Zoom, streaming services like Netflix, Twitch, and near-instant communication without lag are nice to have, but are they essential? No. For most of us, if we are honest with ourselves, having to wait an extra few seconds before a post can load onto our screen is irritating, but overall it is not really hindering our lives. These satellites should be reserved for the people on the planet that need them most, where the most benefit will be felt, economically and socially. Will a multi-billionaire business owner choose societal stability and progress over profit? I think we know the answer to that, but I'm ready to be proven wrong.

I do concede that Musk has sent satellites to places in need, such as Ukraine, where they have, literally, been a saving grace in the war against Russia. On 24 February

2022, a mere hour before the first missiles announced to the world that Ukraine was being invaded by its bull-ish next-door neighbour, the Russian military launched a cyber-attack, hacking the American owned Viosat satellite, utilized by the Ukranian military for communications. The Ukrainian military, at its most critical hour, was silenced from behind a screen somewhere deep in Russia. Unable to talk to front-line troops, civilians were also left scrambling for connections to networks for vital news and information on how to get to safety.

In little under a week, Starlink satellites were in position and the hardware fired up. From drones dropping bombs to an actual lifeline to the outside world, Starlink foiled Putin in a way he did not foresee. He could not silence Ukraine. He has struggled to keep Russian citizens unaware of the world's reaction to the invasion. It has empowered Zelensky to bring the world's attention to Ukraine, and Twitter has meant he literally has the world's politicians in his pocket.

Understandably, the eyes of the world's military focused on Starlink technology. Not only were the satellites perceived (somewhat incorrectly) as an act of life-saving charity[¶], it was also one hell of a PR campaign for the company.

[¶] Important to note Starlink haven't provided all the infrastructure and internet for free. Donors and sponsors have come from many sources, including a huge amount from US Government and USAID.

We are very much heading into Stark Industries territory now. Of course, these satellites have their benefits, but will the Western world be so keen when China and Russia develop their own mega constellations?

NASA, scientists, and organizations around the world have expressed concern about the sheer volume of satellites being launched. Between 1957 and May 2020, we had placed 2,200 active satellites. In the two years following, that figure doubled. Currently there are around 25,000 trackable objects, 5,000 of those being satellites. The SpaceX expansion alone will see that number more than double.

These launches have pioneered a way for multiple space prospectors to mine their own fortunes in space. Amazon's Project Kuiper, Astra, Telesat and the UK-based OneWeb all have plans to launch their own 'mega constellations'. The LEO is already crowded, and as more of these companies clamour to become competitive in this realm, at this current expansion rate, we will see satellite numbers in the hundreds of thousands by the end of the decade.

These satellites, like many things discussed in this book, are increasingly threatening Dark Skies. The light pollution caused by them impacts astronomers, those who study space and defend Earth and those who simply enjoy stargazing. Research by the International Asteroid Warning Network, which looks for potential asteroid threats, is being increasingly blighted by satellites that can be difficult to tell apart from actual asteroids. Not being able to see the wood for the trees is one thing, but the asteroid for the satellites? This is a genuine risk to all life on Earth.

The night sky belongs to everyone. It is a shared global heritage. One company, in one country, should not be given authorization to launch objects into space that will impact every single person on the globe and their view and connection to the night sky. It should be a global consensus. We are witnessing some form of space colonialism and compounding the monopoly of a few rich people who will hold the world's communication in their palms. The first ones to get there beat the rest of the gold rush, and charge everyone else a premium for use of their satellites. Despite impacts for every one of us, we aren't getting a single say in what happens. This is the Wild West . . . in space.

For many cultures, the stars are an integral part of life and still serve a purpose. To have this erased, altered and impacted so heavily in such a short span of time by light pollution from satellites is heartbreaking to the many indigenous communities that these satellite companies claim to want to support. Voices from these communities aren't just being marginalized, they are being completely silenced.

Dr Aparna Venkatesan is a cosmological force for good in astronomical science and has fought hard for the inclusion of indigenous people and women in astronomy.

The first woman to ever receive a degree in astronomy from Cornell University, she is now a cosmologist at the University of San Francisco. She holds the unsettling view that by 2025, the night skies, even the moon, will be permanently altered for all of us on this planet that we call home. To let that permeate properly, the night sky has remained wholly unchanged since the beginning of human existence, up until the last fifty years. Venkatesan recommends that humans need to radically shift their understanding of space, and instead recognize 'the view of space as an ancestral global commons that contains the heritage and future of humanity's scientific and cultural practices'.[2] I wholeheartedly agree. Many conversations must be held and solutions found to the problems posed by satellites. To allow launches to continue before solutions are found will be an expensive mistake, and one we will all pay for.

Various communities are increasingly nervous about what this means for the LEO, for humans, for nighttime on Earth, for our wildlife, which depends on it. The International Astronomical Union has brought these voices together to create the 'Centre for the Protection of the Dark and Quiet Sky from Satellite Constellation Interference', launched in June 2022. This brings voices together from the worlds of science, ecology and indigenous communities and aims to raise awareness, pushing for regulation of the satellite industry. The United Nations Office for Outer Space Affairs is also heavily involved.

Yes, there is a need for satellites, but we cannot allow our night skies to be forever changed at frightening speed without consideration of the impacts. Space itself is a fragile environment and ecosystem that we must protect. Sustainability and protection of this unique ecosystem must be a core requirement for all companies launching satellites. The current rate at which we are launching satellites is completely unsustainable.

What goes up, must come down. Many of these satellites will burn up on re-entry to Earth's atmosphere but others float on, going to the great space junk graveyard in the sky, hampering space-exploration missions for future generations. The ones that turn into man-made meteors don't just simply disappear; the chemicals they contain are released into Earth's atmosphere. Some of these deplete our ozone. *Yes, we've been here before, haven't we?* Currently, about eighty tons of junk rains back on us each year, with negligible impacts. But as satellite numbers rise exponentially, this won't be the case for long.

There have already been recorded incidences of the Chinese space station needing to make emergency manoeuvres to avoid a collision with a Starlink satellite. The more that are launched, the more this risk increases. According to Hugh Lewis, head of the Astronautics Research Group at the University of Southampton, Starlink satellites alone are involved with fifty per cent of all close encounters between two spacecraft, which roughly averages at 1,600 close calls a fortnight.

Due to different regulations in different countries, what is acceptable in the US under the Federal Communications Commission (FCC), would not necessarily stand up to scrutiny under the UK Space Agency. There is a need for global regulation of satellites and management of the LEO, just like we managed with the ISS. There is a very real risk of human life being lost in the near future if not. We should not wait for a tragic accident to take place before regulation happens.

As the rich on Planet Earth seem hell-bent on fuelling the demise of our climate, we need desperately to reduce our fossil-fuel usage. Earth's resources are not infinite, and supplies are running low. Growing wealth disparity means there is a shrinking percentage of the population able to buy fuel for things like heating, cooking and travel. Whereas the mega rich, a handful of powerful men, hoard more than their fair share of our resources, using it to literally launch ways to make themselves even more money, or joyrides into orbit. Is this a good use of Earth's collective resources? Should we not be focusing all that money and energy into fixing the issues here at home?

These launches are like throwing fat into the fire. Each Falcon 9 rocket launch by SpaceX releases 336,552 kg of CO_2 into the atmosphere.[3] To offset this, you would need to plant 245,718 trees per launch. Virgin Galactic produced around the same when sending Richard Branson off to realize his space-cowboy dreams. These figures are cautionary; getting reliable and honest figures out of the private space sector is difficult. Currently, rocket emissions are largely unregulated so we cannot complete the picture. We also must ensure we don't fall into the trap of focusing purely on carbon. Both NASA and Jeff Bezos' Blue Origin spacecrafts use liquid hydrogen, which expels no carbon (though its creation is carbon heavy). Could this be the answer, switching to hydrogen for fuel? Unfortunately, as we are increasingly finding with our energy needs, it just isn't that simple.

Since 1965, the Russian Proton programme has launched hundreds of commercial and government satellites using a fuel called Unsymmetrical Dimethylhydrazine (UDMH). It is highly carcinogenic for humans. The world's oldest spaceport, Baikonur Cosmodrome, sits within the globally important ecoregion of the Kazakh Steppe.[4] At 2,000 km, it stretches from Kazakhstan into Russia, from the Caspian Sea to the Altai Mountains. At the crossroads of the Central-Asian and Siberian-South-European flyways, it hosts millions of migrating birds on their travels, alongside the grey wolf, Eurasian lynx, and the wonderfully weird but critically endangered saiga antelope. Rockets launched from Baikonur leave a trail of chemicals that return back to Earth, sprinkled liberally into the air.

Soviet scientists have dubbed UDMH 'the Devil's venom'. The fallout caused by the pollutant seeping into soils has been devastating; a large area of the Steppe is now dead; an ecological wasteland.[5] Studies have linked spills from rocket mishaps and particles from launches** to increased rates of rare cancers in nearby areas of both Kazakhstan and Russia. News broke recently that there have been at least ten incidents of spillages of UDMH into the Barents Sea and Pikialasorsuaq (North Water Polynya) since 2002.[6]

Pikialasorsuaq is one of the most ecologically important areas of the Arctic, with iconic species such as narwhal, beluga whale and walrus all depending on its open water. Encapsulated by ice on all sides, this area of open sea is vital to their existence. The Inuit populations of Canada and Greenland did not gain information of these spillages until 2017, causing considerable distress to these communities, who are equally as dependant on these waters and its wildlife for survival. These actions disproportionately leave them exposed to severe health risks, and breaks a

** Including launches paid for by the European Space Agency.

number of laws and treaties regarding space, environment and health.

In 2020, the UK Government paid $500 million of the taxes that you and I pay, to own a thirty-three per cent stake[††] in the UK-owned company OneWeb, saving it from bankruptcy.

OneWeb were utilizing the Russian cosmodrome to launch their satellites, including launches powered (in part) by UDMH.[7] OneWeb launches from Russia have, as of 4 March 2022, been grounded. ROSCOSMOS removed a rocket from the launchpad that was carrying thirty-six OneWeb satellites, after demands of the launch – that the UK Government disinvest their share and prove the satellites wouldn't be used for military reasons – were not reached. New rocket fuels are desperately being researched to replace these dangerous and dirty methods, yet they are still being used by European, Russian, Indian, and Chinese rocket designs, with particles spilling from various rocket stages, or being encapsulated to float in space as highly toxic waste.[8]

The scientific community admits more research is needed into the impacts of space exploration and the burgeoning business sector that is space tourism and satellites. We don't yet understand the implications of our

[††] Later dropped to nineteen per cent.

advancements to human life, the climate, our environment, biodiversity and of course our night sky, which is already infiltrated by hundreds of satellites, ruining the experiences and connections of many who look up. I dread to think that we could have a night sky we do not recognize in just a few years' time.

It is not too late to intervene and there is too much at stake. We cannot afford not to press pause on launches until there's a global agreement on satellites and their usage. We must learn from our mistakes on this planet before we go destroying and altering others. All voices must be listened to in the debate and answers must be first found to things like light pollution from satellites and the impact of rocket launches on our climate. We are creating yet another mess we don't know how to clear up and must learn that prevention is better than cure. The profits of a few must not dictate the consequences for the many. If nothing else – think of the poor people staring at dots on screens at RAF Fylingdales. They've got enough to keep an eye on.

CHAPTER TEN

How We Can Fight for the Night

I am completely and utterly lost. Unzipping my backpack to remove my camera, the harsh metal sound seems unnatural, so out of place, surely it cannot always sound that loud? After that there is no human sound. No hum of electricity from a pylon, nor rubber on tarmac from roads. Not even my footsteps on the soft and springy shore. Instead, the gentle sea breeze carries with it the sorrowful song of grey seals, and the gentle rippling tide against storm-smoothed pebbles, each exhale gently releasing the susurrus from the spindly coastal grasses. The seals gently sing, their voices ghostly, carried from a cove invisible to me on the grassy edges of the island. Their short sharp breaths punctuating the ebb and flow of the tidal race. Above me is a glittering canopy of thousands of twinkling lights. I search for familiar shapes, the stories I know in the darkness that root me to my place on the ground.

I am caught by surprise that I struggle to wade through this vast ocean of never-ending star light, so many that I cannot find my way markers. I have never experienced anything like this. It takes much longer than usual to find my familiars up there, my celestial cairns and beacons. They're not used to the competition of their neighbouring stars. Usually by far the brightest to the naked eye, they find themselves less conspicuous, camouflaged in the cosmos. Two distinct cloudy bands arc across the sky, cascading like white water into the sea on the horizon. I have never been happier to be lost. A red beam sweeps the land in front of me, briefly revealing the shape of the shore before it circles away, smoothing over the ridges and crags before reaching beyond the land's edge and skims the coalface of the Stygian sea, reaching out across the darkness, raising its warning.

Ynys Enlli has long been a lodestar to the people that came before us; the saints and the sailors, the invaders from distant shores. Before them, to the wildlife, the birds that for centuries have flocked to the sanctuary of Enlli on this western edge of Cymru. Its iconic lighthouse with its square, red-and-white striped tower reaching into the sky, is a beloved landmark to all lucky enough to find their feet on the shores of Enlli. I walk with the lighthouse behind me, the large sweeping beams somewhat unsettling, completely silent, as they continue their ritualistic revolutions of warning. Perched on the south-western edge of the small island, it has stood as a beacon to sailors in the night since 1821, signalling the salt dogs to keep clear of its rocky shores. As I walk, the presence of the light feels like a physical force. I want to shrink away as the beam passes by.

Another sound travels through the night air, a strange call I had not heard before, similar to a rooster that can't quite work out how to crow. I realize my path is blocked by a small, shuffling pebble. I switch my headtorch on at its weakest setting so as not to startle this peculiarly animated rock. In the dim light, I see two eyes sparkle back at me. Totally unperturbed and in no real hurry, I realize I've stumbled upon a Manx shearwater. A squat little black-and-white seabird when seen on the ground, its inquisitive eyes shine out from its head like two pieces of polished jet, looking directly into mine. I thought it would have immediately scarpered when faced with a human, but it actually walks towards me. Then, as my head torch scans the area, I realize it isn't alone: there are a handful more, like tumbled stones on a shore. To see them shuffling, almost clown-like, belies their true nature.

When seen out at sea, that's to see them in their element, their powerful wings beating as they shear across the face of waves, their legs built for swimming and life at sea, definitely not built for a life ashore. Unfortunately, these birds haven't figured out a way to have nests at sea, so must return, every year, to the islands they were born on, where they roost in burrows underground, like feathered rabbits.

I thought I was too late in the year to witness the famous Manxies of Enlli, who raft in their thousands on the sea around the island until darkness falls. They will only return during the darkest of nights, using the darkness to protect them from land predators. As darkness falls, they take to the skies, descending on the island of Enlli. Home to over 20,000 pairs of birds, the noise is said to be spectacular. Once believed by sailors to be the calls of witches or pirates at sea, it fills the night air. They apparently take no prisoners on landing, either; many have told me they've been a crash-landing pad for a Manx shearwater. I must be witnessing the stragglers, the final few left to fledge who have yet to launch themselves out on their maiden voyage, a 6,000-mile migration, non-stop to Argentina.

It's an incredible journey, one that must be done alone with no parent to guide them. These birds, their lives, depend completely on naturally Dark Skies. Without darkness, they cannot orientate themselves when they fledge. On their first flight, newly emerged from their burrows, the Manx shearwaters must find the natural glow of the horizon and use the stars to find their way. However, humans have now filled the horizon in every direction with artificial light, the light we produce spilling upwards, obscuring the natural cues the young bird instinctively looks for. Instead they are drawn off course and inland to towns and cities.

Enlli is a haven for humans and wildlife alike, but its lighthouse, that proud fixture on the skyline saving lives at sea, was once murderous. When the light was electrified, the lighthouse, instead of signalling danger, became an irresistible siren for a multitude of birds, drawing them in, by the thousands, to their deaths.[1] For decades, the wardens and ornithologists at the island's bird observatory tried to figure out why the lighthouse was causing birds to flock around it, some to their demise as they either collided with the structure or flew around it and eventually died of exhaustion.

A warden who was there in the late 1960s told me how distressing it was to find and collect each morning the injured birds and the bodies of those less fortunate birds. The warden, Hugh Miles, was kind enough to send me some extraordinary photos that he took at the time, the sweeping beams becoming trawlers nets, the images showing tens of birds seemingly trapped in the bright-white beams of light. Then in 2014, the colour of the lighthouse lamp changed. Overnight the bird attractions and fatalities stopped. That's why if you visit today, you'll see the lighthouse shines red, not white. An incredible example of the power light can have, and the destruction it can cause. Now the lighthouse stands as a warning, not just for those at sea, but also to us of the problems of light pollution and its impacts on wildlife.

A sanctuary for its 20,000 saints, the sailors who took refuge there and for people who want to be immersed in nature today, it was a moment of national pride in February 2023, at the end of Welsh Dark Skies Week, when Ynys Enlli became Europe's first and only International Dark Sky Sanctuary. A beacon to all who love and care for our Dark Skies. Thanks to the hard work of the island's community, which started with the Porter family and was carried over the finish line by warden, Mari Huws, the sky above Enlli will be protected forever more.

Dark Skies is a movement studded with its own stars, beacons of darkness who burn brightly to warn the world of what our light addiction costs.

We cannot talk about Dark Skies in the UK without mentioning Bob Mizon. To many he is viewed as the grandfather of the movement and he has campaigned relentlessly to open our eyes to the destruction and losses. His journey to darkness started in post-war Dagenham, London. As a young boy he remembers playing outdoors of an evening with his friends and being able to see the night sky in all its glory from London's East End.

'This was only possible because World War Two had bankrupted us as a country, we couldn't afford to have the street lights on past 10 p.m. so London had a Dark Sky! We just used to go out and look up at the stars, it was wonderful.' But eventually, the lights came back, and the stars were lost to the Londoners beneath them.

Since then, Bob has seen us succumb to the irresistible lure of excessive lighting, witnessing how much we have lost. He has not just watched the rise of the Dark Skies movement but been central to it. Joining the British Astronomical Association in the 1970s, members realized the skies were disappearing and knew that if astronomy in Britain had any hope of survival, they must do something about it. And do something about it they did. In 1989, they created the Committee for Dark Skies,* and together they rallied, lobbied, and fought tooth and nail for their cause. Bob says it wasn't easy – 'They thought we were a group of people throwing bricks at lights or something.' But they fought on, joining forces with the Campaign for Rural England (CPRE) to educate local authorities and highways agencies. They secured a major victory in the 1990s, when they convinced the UK Highways Agency to only use fully cut-off lighting.

Bob is optimistic about the future of Dark Skies and thinks we are on the right path, but a lot more work needs to be done for people to understand that LEDs, while energy efficient and cheap to run, are currently always too bright and too blue. But he has seen changes, not just in lighting, but in people's attitudes.

* Now the 'Commission for Dark Skies'.

For over three decades he has raised awareness about a problem no one knew existed. It hasn't been easy. Over the years he's received hate mail and abuse from those who misunderstood what he was trying to achieve. But recently, he was giving a talk in a primary school and asked the children what stops them seeing stars. 'A little girl in the front row stuck her hand up and I was expecting her to say clouds or something, as they usually do, but when she answered, she said light pollution! I thought blimey, the message is getting through,' he told me. I asked Bob what's kept him going all this time to fight for the night, and it's now that his voice cracks, the raw emotion breaking through. 'If my granddaughter can see more stars than I ever could in East London then I'll be happy. It'll have all been worth it.'

It's through Bob Mizon that I became aware of Joy Griffiths. He describes her as a 'tiny woman and a tremendous campaigner'. Joy volunteered throughout her life to raise awareness about light pollution and influence change. Out of the thousands of members of the British Astronomical Association, just over a hundred volunteered to be local representatives to raise light-pollution awareness. They would deliver leaflets, run local engagement events and, crucially, trawl planning applications, pushing for lighting policies and for the nightscape to be considered in the planning process. The Dark Skies community sadly lost Joy at the age of 53 in 2006, but her legacy lives on through the annual Joy Griffiths Award, given to those for 'meritorious efforts in the cause of darker skies'.

In 2010, the Joy Griffiths Award recipient was the very deserving Emma Marrington. Emma is the brains behind the CPRE's annual 'Star Count'– a campaign encouraging people across the UK to pay attention to their night skies. It can be done by anyone, and needs no equipment. You simply have to go outside during Star Count week (usually over the new moon in February) and count how many stars you can see in the constellation of Orion. You don't even need to be in a Dark Skies location, as the beauty of this campaign is that it's a citizen's science project, designed to get a picture of light pollution from the ground around the country. This is vital, as satellite imagery of light pollution doesn't tell the whole story. You simply count the stars and submit them to the CPRE Star Count website, which you'll find in the back of this book.

Since its inception in 2006, thousands of people have taken part, adding to the CPRE's 'Night Blight Map'.[2] Emma remembers her first experience of Dark Skies being one that was shocking to her as a teen:

> *I grew up very close to Heathrow Airport, so I saw no stars at all growing up. Mum used to take me youth hostelling around the UK, and one year we went to the 'Heartbeat Lands', the North York Moors National Park. We got the train to Goathland and walked to a hostel in the middle of*

> *nowhere. I remember waking up in the night and looking out of the window and being in total shock at the stars. I remember these truly, really Dark Skies initially being quite scary, for a kid from a very urban location I just wasn't used to it.*

To Emma, that's what makes Star Count so special, she adds:

> *Many people won't have even thought about the night sky, especially in urban areas. But until you're given that opportunity to look up and take notice, you just don't know what you're missing. The worst I've counted is just nine stars where I am now, and a miniscule amount of people see over thirty.*

She's right. Urban areas need to be given the opportunity to look up. It would take some minor changes to lighting on the ground that would enable more urban populations to enjoy the stars. Even in our most light-polluted areas in the UK, many have little dark pockets. Okay, so they may not be incredibly dark, but it's about relative darkness; how getting to a place even marginally darker than your street will positively impact your mental health and wellbeing and reconnect you with our night skies. Emma tells me of an urban park in Kingston:

> *I go to Richmond Park in London of an evening and when you get away from the lights it makes such a difference. The*

street lights here come right up to the park's boundary, but inside there's no lights. It's such a transition. It may not be the deepest of Dark Skies but my eyes still need to adjust, I can see more stars and I get to experience the sound of owls and other wildlife. It's a little oasis of darkness, and these are so important to people living in towns and cities.

We should be actively reducing the light pollution around these little oases of darkness, to protect them as a refuge for humans and wildlife alike in our ever-brightening cities. By adopting just a few simple lighting policies and, crucially, not being afraid to enforce them, local authorities and communities could instantly gain darker areas, enriching their lives and neighbourhoods with a gift beyond measure – the night sky.

You wouldn't dream that London would be home to a hotbed of Dark Skies talent, but alongside Emma, London is also home to the supremely talented lighting designer, Kerem Asfuroglu. His company, Dark Source, has won myriad awards for its innovative and environmentally sound lighting designs, all of which centre around an ethos of protecting darkness.

Instead of following the norm, Kerem has blazed his own trail, becoming a bright spark in the Dark Skies world:

My response was to do something that matters. My knowledge could apply to this field and make a difference rather than just being a visual aid for wealthy people.

There's a certain irony that it's the light that brought Kerem to see darkness differently:

> *Through working with light I gained an appreciation of darkness. Culturally we talk about brightness, but we must balance brightness and darkness to talk about light. Darkness is the base layer, the canvas. How much you skilfully craft it and preserve it is how you create light.*

I found Kerem through his work on a Dark Sky Community in Presteigne, Cymru, where Kerem has been carefully and considerately working alongside the community to create an entire town that is Dark Sky friendly. I've worked alongside him on projects, and I've always been struck by how skilfully he guides community members to empower them to do what's right with their light. He's not afraid to tackle the commercial side, either:

> *I'm always surprised at how positive commercial clients are. On an NHS project, we looked at light use through COVID, where monuments were floodlit blue. We think it's beautiful, that light became a symbol of life and celebration, but it shows how little we understand of light's negatives. They were like, 'okay, we get it. We don't want floods or uplighting, we want to support Dark Skies'. People will always surprise you if you make them part of the journey.*

Originally from, Antakya, Turkey, community is something that has weighed heavy on Kerem's heart; living through the horrific aftermath of the 2023 Turkey earthquake and having to watch from afar as his hometown was changed forever:

My hometown was a beautiful and historic place, a eutopia of different ethnicities living together in a harmonized way. I yearn for what's gone, the beautiful historic town, the architecture, but even if they built that back, the community is gone. That's where and what a place is. Culture is community.

I am grateful that Kerem is part of our Dark Sky community, and I get to work alongside him in Cymru. 'Hiraeth'. That Welsh word. Mourning for a place that doesn't exist any more. 'I feel that now, I know it.' We both know hiraeth applies to our night skies. But unlike the tragic events in Turkey and beyond, the solution to light pollution is an easy one to fix.

We have encountered issues before, like the ozone layer, and been quick at improving things, I will stay optimistic that we can overcome it. It is the easiest pollution to resolve. Switch it off and it's gone.

Kerem has also been working across the Irish sea, alongside the brightest star in the Irish sky, Georgia McMillan, where they have both been working with Mayo County Council to make Newport the first Dark Sky town in Ireland.

Georgia is Mayo's Dark Sky Park Development Officer who is doing all she can to restore darkness to the Wild Atlantic Skies above Ireland's west coast and its Wild Nephin National Park. She is one of the voices of a fierce community group – 'Friends of Mayo Dark Skies' who are passionate about reclaiming the real jewels of the Emerald Isle, *'We are very much community led, that's where we are strongest,'* Georgia says of the group who are behind the wildly successful Mayo Dark Sky Festival. *'Our festival is community led and people really embrace it. We have found people are ready to change and people are getting so sick of light. We need to encourage them to make official complaints, or it's hard to get the voices heard.'*

A fellow night owl, Georgia grew up in London, combining the cloaked world of stage shows, and the Dark Skies of her mother's Mayo, finding her love for the Theatre of the Night:

> *My dad was a magician, so we were night owls, going to night events more than daytime. Dad gave me a little Patrick Moore book when I was younger. I'm not really an astronomer, I'm more a stargazer, but just like you don't have to know the name of every flower in the garden to appreciate them, it's the same with stars in the sky. Magicians and astronomers are similar beasts. In the magic circle, the darkness is exciting, attractive, and then you have the theatrics of the night sky.*

Much like my experiences across the sea, Georgia used to see people concerned that Dark Skies were sending Ireland backward. It was the late 1970s before they got electricity in some of these communities, but now people are a lot more understanding of what they want to achieve, due in no small part to the tireless efforts of Georgia and the group, '*We need to start looking at ways forward, to get off the grid, it's not backward. It empowers communities and is economic progression,*' she says.

Thanks to the efforts of the Friends of Mayo Dark Skies group, and their partnership with its national park, Mayo became Ireland's first International Dark Sky Park in 2016, creating an area of over 15,000 hectares, which is now protected for the communities that live under them and the future generations. According to Georgia, '*The communities are so vital, if they aren't engaged with landscape and preserving the landscape, they won't care for it.*'

We desperately need to adopt national policies that will address light pollution and recognize it is as serious as water or air pollution. Someone working hard to get the light-pollution problem under the noses of our politicians is David Smith – Social Change and Advocacy Officer for the charity, Buglife. This is not his first rodeo. David is behind some of the most powerful environmental campaigns of recent years, including the Plastic Free Communities and

Safer Seas Service for Surfers Against Sewage. We're lucky to have him on Team Darkness.

Buglife has long been aware of the impact of light on our invertebrates, with their review in 2011 raising the alarm; but it mainly fell on deaf ears, despite having a plethora of evidence pointing to light pollution as one of the main drivers of the insect apocalypse'.[3]

When he first came to the dark side[†], David was surprised to find that despite having a very large evidence base, there is a distinct lack of non-government organizations (NGOs) working to combat the issue:

> *I expected to see a large group of environmental NGOs working on the issues, a collective group who could coordinate and help drive the issue forward. However, that has not transpired and there is a large gap in the environmental sector for work on light pollution, especially those looking at species.*

David is working hard to untangle the web that is the parliamentary process, finding gaps through which to push light pollution up the political agenda. His aim is to create policies and targets that we can then hold the government accountable for:

[†] And 'The Force' is definitely strong with this one.

> *We need to recognize light as an environmental pollutant. Once it is recognized as a contributing factor, it will increase the priority of it, and we should see strategies and targets to measure in place that are enforceable. This should lead to change. Currently there is no active work stream in DEFRA on light pollution. Scotland are looking at introducing light-pollution measures within their biodiversity plan, and Northern Ireland are adding it to their bill. The EU zero-pollution action plan didn't include light but now light pollution is considered within that – that's a big change and we could potentially see the first EU wide light pollution target.*

In Westminster, we did see the All Party Parliamentary Group for Dark Skies form in 2020, but, after a promising start, that seems to have ground to a halt.

David agrees that we need to act fast, to remove a huge environmental pressure for our vulnerable insects and other wildlife that we can fix quickly and cheaply:

> *There needs to be much greater action across the whole environmental sector. It needs to move quicker. If we can address something we fundamentally know how to do cheaply, it will buy us more time everywhere else, it shouldn't be put to one side because there are other issues.*

David has fond memories of one charismatic insect that is suffering heavily thanks to ALAN – our glow-worms:

> *I remember France in the middle of the countryside, camping, and there being glow-worms, everywhere. I can vividly remember that, I was in my early teens, I went from never seeing one to holding one in my hand. I've never seen one since. One problem is 100 per cent linked to light. I would love to be able to see that again. You could see all the stars, the constellations, all the things I couldn't see at home in Hampshire, it was an eye-opening experience that I still reflect on.*

You'll be forgiven for not realizing we even had glow-worms in the UK. I'm told reliably that these enigmatic creatures once adorned our countryside in their thousands; glorious baubles in the summer hedgerows. They are extremely sensitive to light, shying away from anything over 0.5 lux ('lux' being a unit used to measure the intensity of light hitting a surface). This is the reason, despite spending most of my life outdoors at night, that I have yet to see a glow-worm in real life. In England alone, we have lost seventy-five per cent of our glow-worms since 2001, within just two decades. Despite my best efforts, there's a very real risk that I will never see one. They could be wiped out in my lifetime.

Dark Skies is about so much more than astronomy, it's about darkness in the environment, or as it's known in the field: the 'Dark Ecological Network'. *Why aren't we acting on this now when we have the solutions?* What we are missing in our movement is the visible consequences of impact. Although devastating, it goes by mainly unnoticed by the average person. We are missing our turtle with a plastic straw stuck up its nose. The social-media clip that forced the hand of the world into action against plastics.

It's easy to fall foul of greenwashing, with claims of efficiency and climate-friendly solutions hiding the truth of their impacts to our natural world. As David Smith says, 'LEDs are a greenwashing tool that are intrinsically linked to the lighting companies. The lighting industry is marking its own homework and the government are happy for them to do that.'

So here is my gauntlet, laid at the feet of the lighting industry: if you build it, they will come. Your profits needn't suffer in the name of darkness. There's a burgeoning market, just ripe for the taking, if only you would reach up and take it. The change is coming, the consumer who wants to take care of the world and the policies that will make the changes happen, they're on the horizon. The only question is, which company will get there first.

Which brings me to Andrew Bissell, another wonderful lighting designer and long-time sufferer of my ambitious plans for darkness. He first stumbled into darkness when the company he worked for won the contract for the Northumberland National Park's visitor centre, 'The Sill', in 2014, which was designed with darkness in mind.

Before then he thought Dark Skies were mainly astronomy focused, but realized the principles should be used everywhere:

> *We've always been careful with light pollution, adhering to ILP[‡] standards, which are not quite Dark Sky, but this design had to be a higher standard. We started to think about internal lighting and its impacts, proposing motorized blinds that came down at dusk as the building is heavily glazed. I realized why aren't we doing this everywhere? All the things we have done here could be done everywhere and we could make everywhere better. That's when I went back to the IDA and looked at how do we get this everywhere, how do we push it forwards.*

Growing up in Cymru, Andrew was accustomed to the dark:

> *I just accepted it was dark. My first proper memory was Tottington, not far from Manchester and it was the lack of Dark Skies. Looking up and seeing half a dozen stars and it dawns on you that you used to see thousands.*

Andrew knows how small the Dark Skies world was, too, but people are starting to embrace darkness, in all walks of life:

‡ Institute of Lighting Professionals.

> *Suddenly everyone is talking about it, it was a lonely crusade at first, but now people are disappointed if they arrive expecting darkness and they're looking at car park lighting. Big online retailer warehouses, offices, data centres, hotels, they're all up for it. Retail is lagging, they still see direct financial impact in their mind that the brighter they are the more they will sell, but they will realize this isn't the case.*

Andrew wants to see an introduction of a Dark Sky building certificate, believing it would help create an impact in cities where everything is about the competitive edge.

> *When we first started on projects, getting Dark Sky lights was difficult, but it is getting easier as companies realize that they are losing out. Lighting manufacturers just want sales, we are starting to get it, but missing a trick.*

It is easy to become an advocate for Dark Skies. First, ask yourself, have you ever seen the stars? Throughout my time as a Dark Sky officer, I have come to realize that more often than not, the answer is no.

Secondly, ask yourself, why haven't I seen the stars? Have you simply not taken the time to step outside and look up? In that case, that's easy to solve. Step outside and look up.

Then you'll want to know how dark your sky is. If you can't see any stars at home, then you are living under a heavily polluted sky. If you can see some, why not try out the CPRE Star Count technique? Look for the three stars of Orion's Belt. Once you have those in your sights, see how many other stars you can see in this constellation. If you can count fifteen stars or fewer, you are in a very light-polluted sky. Identifying more than thirty stars means that you are part of the two per cent of the population living somewhere with truly Dark Skies.

Whether you are in a Dark Sky area or not, there are things we can all do to ensure we are doing everything we can to reduce the impacts of artificial light at night. Remember it's not just about stars. Light at night is impacting our health, that of our communities, and seriously, our biodiversity.

Look at your home and think about your light use. Are you a light polluter? Most of us are, without realizing. All lights have an impact, from super powered flood-lights to tiny, garden fairy lights that find themselves woven round trees and trellises. Every bulb, big or small, impacts the night around it.

What lights on your home are really needed? All lights should have a clear purpose ('looking pretty' is not a purpose). Most of us spend very little time outside at night and have lights installed that are more powerful than they need to be. Unless you're operating heavy machinery in the middle of the night, you don't need flood-lights. Sometimes, a suggestion of light is enough to show you where to go. If the lights aren't needed, switch them off. Make sure the light doesn't have the dusk till dawn sensors, and use with a motion sensor if needed. Motion sensors work particularly well if you want a light to turn on as you pull onto your drive, for example. They'll also be more effective at alerting you to any unexpected movements around your property.

If you genuinely need lights outside to assist you of an evening, there are some simple rules you can follow to make sure they're as Dark Sky friendly as possible. Remember the more lights you put out, the darker everything else around them will seem:

- Make sure the light is fully shielded and downward facing.
- Up-/downlighters are a big no. You want all that light you're paying for to be effective and efficient, any light fitting that allows light to spill sideways and upwards is contributing to light pollution. This also helps to cut out glare, saving your eyeballs pain.
- Use energy-efficient bulbs that are only as bright as needed. Lights that are too powerful will create glare and an area that's over-lit, casting deeper shadows in the area outside the pool of light. This often leads to us putting more and more lights out, trying to chase the shadows away.
- Kelvins: the colour temperature of the light is what really has the potential to create havoc. Bright white

> LEDs are harsh and cold, emitting too much blue-violet light. It's those shorter wavelengths that are rubbish for your health and wellbeing and are the ones that lure insects and other wildlife to their deaths. (To be Dark Sky friendly, use lights that are 2,700–3,000 Kelvin, but the lower the Kelvin, the better for wildlife.)[§]

Finding Dark Sky friendly lights used to be a lot harder but, thankfully, they are increasing in availability; the more people want to buy them, the greater the supply. You can find small selections in most hardware stores, and if you can't see them, ask them why they aren't stocking Dark Sky friendly lights. Our big hardware chains could be doing so much more to enable us to choose Dark Sky friendly lighting, but unfortunately many of the big hitters here in the UK have all remained stubbornly silent when I've requested their help, so please add your voices to mine so they can no longer ignore it.

For those of us living with heavy light pollution, don't give up. There are steps you can take to start reducing light pollution levels where you live.

[§] To double-check the Kelvin rating of your bulbs at home, just look at the packaging that the bulb came in. Nearly all bulbs now come with the Kelvin rating displayed, usually with a colour scale ranging from red to blue, on the box.

First, sort out your own lights. Dark Sky schemes are aesthetically stunning. They create a much more pleasant nocturnal environment and create a cosier, more welcoming atmosphere. No one wants their home to look like a border-control barrier, or a tacky, neon flashing nightmare.

Windows can spill light far and beyond their intended target – the inside of your home. There has been a steady rise in floor-to-ceiling glazing around our most sensitive sites, like national parks and AONBs. Of course, people want to make the most of those sweeping views, the vistas that made them buy the property in the first place. But if you want to live in a protected area, you must become a guardian of its natural landscape. Give back as much as you take from the landscape. Glazing may make use of daylight streaming in during daylight hours, but after dark you become a giant light box. Everything in your home is visible for miles around, including you. So, unless you're an exhibitionist, I don't think this is the intended impact of all that glass. Invest in smart glass, thermochromic panes that change from transparent to opaque, working like a one-way mirror, keeping your privacy intact and stopping pollution in its tracks. Ordinary windows are escape routes, too: conservatories, sky lights and if you're dead fancy, cupolas, all contribute to constellation killing. Blackout blinds will stop that, as well as keeping the warmth inside your home, meaning you'll use less energy on heating, too.

Then, show it off. Talk to your neighbours about your new lights and tell them why you're doing it. If you want to go the whole hog, put a little sign outside that says your home is Dark Sky Certified. Most light polluters really don't know that they're light polluters. Having friendly conversations with people educates them and helps the word spread. In time, you'll start to see the uplighters disappear and the stars start to reappear.

Join a community group or start your own. You might find you're not the only one passionate about saving our skies. We are stronger together, and can reclaim the night sky for ourselves and future generations. Talk about why you want to tackle light pollution, inform others about the hazards, give out information, hold a community star-count or switch-off event and see what you're missing. You might even find your community wants to apply to be an IDA Dark Sky Community, like Moffat, Scotland.

Hold some night walks, if you're scared of the dark, which many of us are, get together with others and go for a night-time stroll. Expose yourself to darkness and refamiliarize yourself with your old friend, the night sky. On these walks listen out for wildlife; pay attention to just how much activity there is, even on urban streets, come sundown.

Speak to your community, parish and town councillors. Let them know you and ALAN are through and that you want to rehabilitate your community's addiction to light. Point out to them the energy, carbon and financial savings alongside the health and biodiversity benefits, and they'll soon be sending that sky-glow packing.

Go to your MSs, MSPs and MPs and tell them that star light – not street light –is the most beneficial to you. Show them that you have the solutions to light pollution and that you won't stop until there's constellations above constituency rooftops once more. Write to them, call them, email them. Stand in their offices dressed as a giant light bulb, day after day, or turn up to their surgeries and shower them with their own campaign leaflets folded into origami stars. They'll soon do anything possible to make you leave them alone. It can seem pointless to contact elected members, but we must remember that we are a democratic society and with that comes power. Elected officials work for us, so let them know that this is something you care passionately about. The more of us who do so, the quicker change will come.

If it's already dark where you live, don't rest on those laurels. Light pollution is coming for your skies too, there's no doubt about it. Get ahead of the sky glow and fight to have your Dark Skies protected, so they will be safe from a light-polluted future. Make sure your local authority has a protection plan for Dark Skies in the form of planning policies for lighting. Prevention is better than cure. By recognizing just how valuable that darkness is, you can ensure it is never diminished.

So, in short, the Dark Sky Friendly rules are:

- Use fully shielded and cut off lights that are downward facing only.
- Use energy-efficient bulbs that are the lowest level of brightness needed for the job.
- Use timers or motion sensors, never dusk till dawn sensors.
- Use lights with a warm colour temperature, 2,700–3,000 Kelvin. The lower the Kelvin, the better for wildlife.
- Use blackout blinds and draw them after dark.
- If it's not useful, switch it off.

Each one of us, no matter where we live, can implement change. Where light pollution is concerned, every bulb counts. If we could all be just a little more considerate with how we light our homes and streets, the combined impacts will be huge. The best thing about light pollution is, when it's gone, it's gone. There's no rattling around the planet for thousands of years like plastics, no slicks to be mopped up like oil spills. This is one of the quickest and easiest wins when it comes to giving our wildlife and planet a fighting chance to come back from the brink. The power isn't just in your hands, it's at your fingertips. With just a flick of a switch, you'll be changing the world. So do not go gentle into that good night; rage, rage against the installation of the light. Rage, rage against the dying of the night.

Our lives depend on darkness.

Acknowledgements

Ryan Phillips, my long-suffering husband who has given up on asking what time I will be home. He has supplied endless caffeinated beverages and made sure I sat down to finish this book every time I thought I couldn't do it. He continually pushes me to achieve things he knows I am capable of even if I don't believe him at the time. He is truly an exceptionally handsome and guiding light (and always helps me find my glasses).

Robert Robertson, my nature wizard of a dad who first showed me the stars and the creatures that live under them. My mum Janet Robertson for always telling us star-filled bedtime stories and putting up with all the sleepless nights that all my siblings and me put her through. Thanks to both of them for being brave enough to take the plunge and move us away from a city life that would have led me down a much different path to a life in a new country and culture that made us wealthy in sea salt and sandy feet (even if I didn't thank you when I was a teenager who thought they were missing out – I soon learned!).

My nan, Mary Ethel Potts, who we sadly lost before this book was published. Sonia Nicholls, who would have found it the most amazing thing ever that I had written a book and would have told everyone she came into contact with about it whether they wanted to hear it or not.

This Phillips family who have given us a home from home in the South.

Bucky and Luna, my shining stars that left my side too soon. Logan, George, Bear and Merlin, my furry tribe and night-time adventure companions.

Jonathan de Peyer from HarperCollins who found me among the weeds and gave me this fantastic opportunity, thank you for seeing the glimmer of what my book could be and for starting me off on this path.

Ben McConnell who has been the best editor a girl could dream of. Thank you for making this process as unintimidating as possible and for herding my words out of rabbit holes and into their places in this book. Putting something creative into the world is terrifying but through enduring my terrible chat about space dogs and attempts at humour, you have settled my nerves and given unwavering support. Even when making brutal decisions that were difficult to hear, you did so with such grace and compassion that it took the sting right out. Thank you so much.

The incredible covers for the hardback and paperback editions of *All Through the Night* were created by the very talented designers Matt Burne and Lucy Sykes-Thompson, respectively. Thank you for doing such a wonderful job.

Until I wrote this book, I had no idea of the number of people it takes to get it from my brain and onto the shelf. I have so much more appreciation now for anyone who works behind the scenes in making books happen. Jo Ireson thank you for being a safe pair of eyes and for being a fellow nocturnal-loving human.

Mike Parker who read my first scribblings and gave me invaluable advice, as a writer I've long looked up to, I was blown away by the time and compassion so readily given, diolch o galon.

Clare Scott, our 'Surfabella' who first sent me the job advert for Dark Skies Officer and forced me to apply when I thought I didn't stand a chance. You well and truly pushed me into the party wave of my life.

Lyndsey Stoodley who has been a constant from surf to starlight in pushing me to write about something, anything and everything. They have pushed me into waves and out of my comfort zone all around the coastline of Cymru and is an exceptional human being of huge heart and mind.

My 'Dark Skies family' who work alongside me to protect the darkness across Cymru; David Shiel, Gwenno Jones and Ceri Lloyd who are true bastions of the East, who have big ideas and aren't afraid to make them happen. Alun Owen, watcher of the Isle, a steadfast friend as well as inspirational change-maker and protector of the natural world. Bleddyn Jones, the mindful keeper of Pen Llŷn and the old ways of our landscape heritage. Last, but not least, Rhys Owen, the mastermind and ultimate shapeshifter of Eryri who goes from baling to board meetings without missing a beat. Without these people the darkness we have remaining would already have gone. Thank you all for seeing the bigger picture and fighting for darkness. I am grateful every day (and night) that you all took a chance on me to do the best job in the world – even if I didn't know what biscuit I most identified with – and for letting me think big.

All the fantastic staff at Eryri National Park and the three Areas of Outstanding Natural Beauty (Ynys Môn, Pen Llŷn, Bryniau Clwyd), especially those who have helped me run events and ruined their sleep schedules to help the project. I am in awe of the brilliant minds I get to work with every day, and I am indebted to each and every one of you who have let me tag along and learn so much. A special mention to Gethin Davies, who did a lot of hard work in getting Eryri it's protected status. You helped keep the Park as dark as your heart. To see so much passion and care for our natural world in my colleagues makes the world seem like a better place.

Dafydd Roberts, elusive and excellent ecologist with mountains of patience, thank you for always taking the time to answer my endless questions, sharing your knowledge and correcting my Cymraeg.

Dr Rachel Harvey, Queen of the Peat, for being an all-round good egg. Stop doing so much for everyone else and learn how bloody brilliant you are.

Lucy Webber, just for being you. Catrin Glyn and Angela Jones for being inspirational women.

All the communities who have welcomed me to speak on Dark Skies and those who have taken it upon themselves to act on it. To every person that has trekked up a hillside with me through miserable cloud for just a glimmer of starlight; those who have nearly frozen solid to listen and learn about the stars; those who send me their amazing astrophotography and let me know about local light polluters, thank you for joining me in the darkness.

To Rob Jones and Hannah Marubbi, a shining example of how to do astronomy in the community. They don't bat an eyelid no matter how wild my requests are, whether that's doing astronomy on a paddleboard or up a mountain. A shoutout, too, to the Xplore! (formerly Techniquest) team in Wrexham who have brought wonder to thousands across North Wales with their portable planetarium.

Sam Ryall who fell in love with darkness and worked hard to ensure Dark Skies got lots of media coverage, including BBC's *Countryfile*, which is what led to Jonathan de Peyer finding me. Without you I would still be talking to myself in the dark.

Dr John Barentine, the absolute giant of all things Dark Skies. A passionate advocate and brilliant scientist who somehow finds the time to speak to everyone who needs his impeccable advice. I believe there is nothing that John doesn't know, and I am always grateful to have his vast knowledge just a click away. He is a real inspiration for us all in Dark Skies.

The International Dark Skies Association who have fought for decades now for Dark Skies across the globe. Their mission is mighty, but they never back down. They're a wonderful team of the sharpest minds and most dedicated darkness defenders. Thank you for supporting our work in Cymru and for fighting for darkness across the world – literally.

Kerem Asfuroglu, a man of too many talents, who patiently takes on my ideas and tiny budgets and delivers beautiful lighting schemes that have won awards the world over. Thank you for never saying no and for always striving to do the best we can possibly do, no matter how challenging the circumstances.

Andrew Bissell, lighting designer and advisor who has endured many late nights and late emails about bad lighting. I have dragged him to every corner of Cymru to look at lights and he has never complained. From standing on an airfield in hurricane winds and rains to blistering sunshine on an exposed farm, he takes it all in his stride with a smile on his face and a love for Cymru. Thank you for all the advice and support throughout the project and this book.

Fiona Collins for sharing the stories of the stars with me and for never ceasing to blow my mind with her knowledge of languages, cultures and the effortless weaving together of myths, legends and landscapes. It is rare to find someone from whom stories sing so effortlessly, reaching over thousands of years and miles to be told in the dark and passed into minds so that the stories live on. Fiona is truly magical.

The Porter family, who perched themselves on Enlli for years and were some of the first to capture and acknowledge the true treasure of the island – its night skies. Without their stories, images and ongoing work, far fewer people would know how crucially important darkness is to this little corner of Cymru.

My cheerleading squad; Charlotte and Maisie Lewis, my adopted Neath family who never said no to an adventure, no matter the weather or time or place; Sarah Tracey, for giving me the tough love when I needed it; and Lucy Nield for always being available for a breakdown. Emily and Becca for being the sanity sounding boards and supporters from afar, sending multiple pet pictures in times of low morale.

Hanna Elin Baguley, Hannah Nel, Kat Lawrenson and Kirk Paton (my 'Astro Coven'). All have braved the darkness with me, felt the fear and gone *Sod it* and done it with bells on. They're a constant source of inspiration and I dream of a day my photography is anywhere near as good as theirs.

Don Cardy, to whom I write an apology rather than a thanks, for I annoy the life out of him with my constant badgering.

My ADHD. I may not be normal, but my ADHD means my curiosity knows no bounds, and I like that.

To all who helped in some way, no matter how small or large, to write this book. I will have forgotten people but know that I truly appreciate every single person who crosses my path and supports darkness in some way.

The Dark Sky movement may be small, but we are mighty, and we are not going anywhere. There are so many others working away whose names should be mentioned, and their work could fill the pages of another ten books: Megan Eaves of Dark Sky London; Ruskin Hartley, Executive Director of the IDA; Dan Oakley, South Downs National Park; Jack Ellerby, Dr Hannah Dalgleish, Brian Heely, Christopher Kyber and Annette Lee.

Finally, a huge thank you to Bob Mizon and his family. There are not words enough to explain the impact Bob had and will continue to have. Bob has left us all heartbroken by unexpectedly leaving this Earth, a few short months ago, to return to his beloved night sky. He was undoubtedly the brightest spark in the long and often lonely night that is the Dark Skies movement. There will never be another like him and I extend my thanks to his wife Pam and their children, whom it was abundantly clear were the biggest stars in Bob's sky. His legacy lives on in each and every one of us and we will continue his fight for the night. The stars' greatest ally has returned home to them, and we will remember him each time we look up.

Useful Organizations and Institutions

The International Dark Sky Association

Home of all things Dark Sky. The International Dark Sky Association (IDA) started in 1988 and is now the recognized global authority on protecting Dark Skies, offering outreach, support and technical lighting advice. Their reach now extends to fifty-one countries around the world. If you want to start on your Dark Sky journey, here's the best place to begin.
https://www.darksky.org/

Prosiect Nos

The North Wales Dark Skies Partnership between Eryri National Park and the three Areas of Outstanding Natural Beauty – Ynys Môn, Pen Llŷn and Bryniau Clwyd a Dyffryn Dyfrdwy. Here is a one-stop shop for all things Dark Skies in North Wales, including the Welsh Dark Skies Week held each February.
https://www.discoveryinthedark.wales/project-nos

Eryri National Park

Home to Cymru's largest International Dark Sky Reserve and the base of operations for Prosiect Nos – The North Wales Dark Skies Partnership. Here you will find advice on where to stargaze, how to become Dark Sky friendly and more about some of the stories in the stars you found in this book. Their website is fully bilingual and is a great place to find out what else is happening in one of the UK's original first three national parks.
https://snowdonia.gov.wales/

Bryniau Clwyd a Dyffryn Dyfrdwy/Clwydian Range and Dee Valley Area of Outstanding Natural Beauty

The Clwydian Range and Dee Valley AONB are part of Prosiect Nos – North Wales Dark Skies Partnership – and run events year-round with their fantastic team of dedicated rangers. You can even join them for a hot chocolate at their shepherd hut on Moel Famau before a Dark Skies walk in winter. Together, the partnership created this easy-to-use guidance on becoming Dark Sky friendly, with a range of examples, from home to barn!
https://tinyurl.com/Dark-Skies-Guidance

Ynys Môn Area of Outstanding Natural Beauty

Another partner of Prosiect Nos, Anglesey AONB hugs nearly the entire island's coastline. The AONB team are

a tenacious bunch, working hard with land owners, home owners and businesses to change lighting to protect the Dark Skies that are at risk of being lost over the island.
https://www.anglesey.gov.wales/en/Residents/Countryside/Areas-of-Outstanding-Natural-Beauty-AONBs/Anglesey-Area-of-Outstanding-Natural-Beauty-AONB.aspx

Pen Llŷn Area of Outstanding Natural Beauty

Home to Europe's only Dark Sky Sanctuary, Ynys Enlli, and the final partner of Prosiect Nos, Pen Llŷn AONB is situated under one of Cymru's Darkest Skies, reaching out towards Ireland. It's one of the darkest places you can get to without getting your feet wet! A small team works hard to keep the 'O' in AONB.
https://www.ahne-llyn-aonb.cymru/Home

Ynys Enlli Trust

Europe's one and only International Dark Sky Sanctuary and home to the fantastic Bird Observatory, Enlli is open to visitors between March and October. You'll need to book an overnight stay to ensure you get to see a magical night sky unfold in front of you, and your only neighbours will be seals and sea birds! Be warned, sometimes you can get there but not back again, if the weather changes.
https://www.bardsey.org/visit

Bardsey Bird Observatory

The Bardsey Bird Observatory has been keeping a keen eye on the sky since it opened in 1953. You can even join them for a night-time walk to visit their famous Manx shearwater.
https://www.bbfo.org.uk/

British Astronomical Association

If you're a budding stargazer but not sure where to start, head over to the British Astronomical Society website where you will find loads of advice. They deliver outreach astronomy sessions and have a monthly 'observing challenge' to get you stuck into space.
https://britastro.org/

Campaign for Dark Skies

A branch of the above, the Campaign for Dark Skies was created in 1989 and has been fighting ever since for better lighting to protect our night skies for astronomy. They have lots of great advice and information on their website on how you can be Dark Skies friendly.
https://britastro.org/dark-skies/cfds_issues.php?topic=about

Natural Resources Wales

Natural Resources Wales are responsible for looking after the environment in Wales. Jill Bullen and her team worked

to create a wonderful map detailing the light pollution in Wales. Want to see how dark it is where you are or where you are going? Head here!
https://luc.maps.arcgis.com/apps/dashboards/1cd6ba8a1d7d4a62aff635cfcbaf4aec

Light Pollution Map

Sadly, not all of us live in Cymru. If you want to see how dark your area is, or in fact, anywhere on the globe, check out this light pollution map.
www.lightpollutionmap.info

Campaign for the Protection of Rural England

The CPRE do all sorts of great work, including their annual Star Count. Here is where you find the details on how to take part each February.
https://www.cpre.org.uk/what-we-care-about/nature-and-landscapes/dark-skies/

Mayo Dark Skies

If you're in Ireland, check in with Georgia and the Mayo Dark Sky Park. They keep their website updated with all upcoming events, in person and online!
https://www.mayodarkskypark.ie/about/our-story

Snowdonia Society

This society worked hard to help monitor the Dark Skies over Eryri in the early stages of the Dark Sky reserve bid. They run lots of volunteering days and are a great way to get out and about in the national park while giving something back to the area. You can become a member on their website.
https://www.snowdonia-society.org.uk/

Dark Source

Kerem Asfuroglu's wonderful lighting-design company. Looking for lighting design that puts community at its centre? You won't go wrong with Kerem.
https://www.dark-source.com/

Ridge and Partners

Andrew Bissell and his team are blazing a trail of darkness with their commercial-lighting schemes. They also work globally, so there's no excuse for not going dark.
https://ridge.co.uk/expertise/lighting-design/

Institute of Lighting Professionals

Find lots of great advice on lighting and find consultants or even lighting courses.
https://theilp.org.uk/

Buglife

Buglife are the awesome charity that works hard to stick up for our invertebrates. Find out how you can help save the small things that run the planet and support this organizations excellent work here.
https://www.buglife.org.uk/

Citation Links

CHAPTER ONE
Who's Afraid of the Dark?

1. https://edition.cnn.com/2017/09/01/health/colorscope-black-fear-of-darkness/index.html#

CHAPTER TWO
This Little Light of Mine

1. https://www.bbc.co.uk/news/science-environment-36492596
2. https://www.exeter.ac.uk/news/homepage/title_877183_en.html
3. https://www.peakdistrict.gov.uk/
4. https://www.cpre.org.uk/wp-content/uploads/2019/11/Night_Blight.pdf
5. Footlight Candle Index, Illuminating Engineering Society
6. https://archive.nytimes.com/www.nytimes.com/books/first/m/maas-sleep.html
7. Night Light Europe https://www.interregeurope.eu/nightlight/

8. https://www.darksky.org/france-light-pollution-law-2018/
9. https://www.bbc.co.uk/news/uk-17665397
10. https://eandt.theiet.org/content/articles/2019/02/uk-councils-ramping-up-led-street light-conversions-to-save-cash-and-carbon-report-finds/
11. https://www.health.harvard.edu/staying-healthy/blue-light-has-a-dark-side
12. https://edition.cnn.com/2019/05/16/hcalth/blue-light-led-health-effects-bn-trnd/index.html
13. https://www.cpresussex.org.uk/news/10-things-you-may-not-know-about-light-pollution/
14. https://www.cpre.org.uk/wp-content/uploads/2019/11/Night_Blight.pdf
15. https://unfccc.int/climate-action/momentum-for-change/activity-database/momentum-for-change-enlighten-initiative

CHAPTER THREE
Away with the Fairies

1. https://thesleepcharity.org.uk/
2. https://www.rand.org/randeurope/research/projects/the-value-of-the-sleep-economy.html

CHAPTER FOUR
A Brief History of Stargazing

1. https://www.jstor.org/stable/2717852
2. https://www.nature.com/articles/nature.2016.19261
3. https://www.himmelsscheibe-erleben.de/en/arche-nebra/verweilen/fundort-himmelsscheibe

4. https://www.english-heritage.org.uk/visit/places/stonehenge/history-and-stories/history/significance
5. Storyland: A New Mythology of Britain, Amy Jeffs (2022)
6. https://www.newgrange.com/newgrange-plans.htm
7. https://www.ed.ac.uk/news/2018/cave-paintings-reveal-use-of-complex-astronomy
8. https://www.newscientist.com/article/2128512-ancient-carvings-show-comet-hit-earth-and-triggered-mini-ice-age/

CHAPTER FIVE
Women of the Night

1. https://theconversation.com/more-lighting-alone-does-not-create-safer-cities-look-at-what-research-with-young-women-tells-us-113359
2. https://reuben.ox.ac.uk/get-home-safe
3. https://www.theatlantic.com/technology/archive/2013/07/night-witches-the-female-fighter-pilots-of-world-war-ii/277779/

CHAPTER SIX
Silent Night

1. British Wildlife, Volume 9, 2014, p.332
2. https://www.bbc.co.uk/news/uk-scotland-glasgow-west-52892194
3. https://www.froglife.org/what-we-do/toads-on-roads/facts-figures/

4. https://eu.boell.org/en/PesticideAtlas-insect-decline
5. https://nycaudubon.org/our-work/conservation/project-safe-flight
6. Initiative 1482-2019/Local Law 15

CHAPTER SEVEN
Celtic Constellations

1. Antiquity, Volume 22, Issue 85, March 1948, pp. 45–6
2. https://www.ucl.ac.uk/lbs/estate/view/2921
3. https://www.nationaltrust.org.uk/visit/wales/penrhyn-castle-and-garden/penrhyn-castle-and-slave-trade-history
4. https://www.welshslate.com/about/our-heritage
5. https://www.library.wales/discover-learn/digital-exhibitions/printed-material/the-welsh-almanac-collection
6. https://www.christies.com/en/lot/lot-4005722
7. https://www.npr.org/2016/02/22/467210492/u-s-navy-brings-back-navigation-by-the-stars-for-officers
8. https://www.coraclesociety.org.uk/partners/carmarthen
9. http://news.bbc.co.uk/1/hi/wales/north_west/2994730.stm
10. https://www.nbcnews.com/news/world/queen-raven-leaves-tower-london-will-kingdom-crumble-n1254296

CHAPTER EIGHT
Under One Sky

1. http://www.hawastsoc.org/deepsky/sco/
2. https://www.gla.ac.uk/news/archiveofnews/2020/june/headline_724839_en.html

3. https://adsabs.harvard.edu/pdf/2020JAHH...23..390G
4. https://phys.org/news/2016-04-ancient-aboriginal-star-australia-highway.html
5. https://nycaudubon.org/our-work/conservation/project-safe-flight
6. https://pib.socioambiental.org/en/Povo:Tukano
7. https://www.gaiaamazonas.org/en/noticias/2019-05-31_maloca-the-big-house-of-the-amazon/
8. https://www.jstor.org/stable/pdf/2800273
9. https://www.icelandtravel.is/attractions/imagine-peace-tower/

CHAPTER NINE
The Final Frontier

1. Prices correct as of August 2022
2. https://www.academia.edu/83213856/The_impact_of_satellite_constellations_on_space_as_an_ancestral_global_commons
3. https://8billiontrees.com/carbon-offsets-credits/carbon-footprint-of-space-travel/
4. https://www.oneearth.org/ecoregions/kazakh-steppe/
5. https://www.bbc.com/future/article/20220713-how-to-make-rocket-launches-less-polluting
6. https://www.highnorthnews.com/en/toxic-rocket-fuel-arctic-waters
7. https://www.nasaspaceflight.com/2021/08/oneweb-9/
8. https://www.chemeurope.com/en/encyclopedia/Unsymmetrical_dimethylhydrazine.html

CHAPTER TEN

How Can We Fight for the Night?

1. Bardsey Bird and Field Observatory Report 2011, Page 122.
2. https://www.nightblight.cpre.org.uk/
3. A Review of the Impact of Artificial Light on Invertebrates, Buglife, C. Bruce-White, M. Shardlow. (2011)

would like to thank the following staff and contributors for their involvement in making this book a reality:

Fionnuala Barrett
Samuel Birkett
Peter Borcsok
Ciara Briggs
Sarah Burke
Matt Burne
Alan Cracknell
Jonathan de Peyer
Anna Derkacz
Tom Dunstan
Kate Elton
Sarah Emsley
Simon Gerratt
Monica Green
Natassa Hadjinicolaou
Jo Ireson
Megan Jones
Jean-Marie Kelly
Taslima Khatun
Sammy Luton
Rachel McCarron
Ben McConnell
Molly McNevin
Alice Murphy-Pyle
Adam Murray
Genevieve Pegg
Agnes Rigou
Florence Shepherd
Eleanor Slater
Emma Sullivan
Lucy Sykes-Thompson
Emily Thomas
Katrina Troy
Daisy Watt